The [illegible]

With Which are Interspersed Some Chapters of Autobiography

Jean-Henri Fabre

Alpha Editions

This edition published in 2022

ISBN : 9789356899063

Design and Setting By
Alpha Editions
www.alphaedis.com
Email - info@alphaedis.com

Contents

TRANSLATOR'S NOTE

The present volume contains all the essays on flies, or Diptera, from the Souvenirs entomologiques, to which I have added, in order to make the dimensions uniform with those of the other volumes of the series, the purely autobiographical essays comprised in the Souvenirs. These essays, though they have no bearing upon the life of the fly, are among the most interesting that Henri Fabre has written and will, I am persuaded, make a special appeal to the reader. The chapter entitled The Caddis Worm has been included as following directly upon The Pond.

Since publishing The Life of the Spider, I was much struck by a passage in Dr. Chalmers Mitchell's stimulating work, The Childhood of Animals, in which the secretary of the Zoological Society of London says: 'I have attempted to avoid the use of terms familiar only to students of zoology and to refrain from anatomical detail, but at the same time to refrain from the irritating habit assuming that my readers have no knowledge, no dictionaries and no other books.'

I began to wonder whether I had gone too far in simplifying the terminology of the Fabre essays and in appending explanatory footnotes to the inevitable number of outlandish names of insects. But my doubts vanished when I thought upon Fabre's own words in the first chapter of this book: 'If I write for men of learning, for philosophers...I write above all things for the young. I want to make them love the natural story which you make them hate; and that is why, while keeping strictly to the domain of truth, I avoid your scientific prose, which too often, alas, seems borrowed from some Iroquois idiom!'

And I can but apologize if I have been too lavish with my notes to this chapter in particular, which introduces to us, as in a sort of litany, a multitude of the insects studied by the author. For the rest, I have continued my system of references to the earlier Fabre books, whether translated by myself or others. Of the following essays, The Harmas has appeared, under another title, in The Daily Mail; The Pond, Industrial Chemistry and the two Chapters on the bluebottle in The English Review; and The Harmas, The Pond and Industrial Chemistry in the New York Bookman. The others are new to England and America, unless any of them should be issued in newspapers or magazines between this date and the publication of the book.

I wish once more to thank Miss Frances Rodwell for her assistance in the details of my work and in the verification of the many references; and my thanks are also due to Mr. Edward Cahen, who has been good enough to revise the two chemistry chapters for me, and to Mr. W. S. Graff Baker, who has performed the same kindly task towards the two chapters entitled Mathematical Memories.—Alexander Teixeira de Mattos. Chelsea, 8 July, 1913.

[Recorder's Note: Most Translator's Footnotes have been omitted from this text, but some of his references to localities and insect names are included in brackets. I apologize to English readers for changes to American spelling.]

CHAPTER I. THE HARMAS

This is what I wished for, hoc erat in votis: a bit of land, oh, not so very large, but fenced in, to avoid the drawbacks of a public way; an abandoned, barren, sun scorched bit of land, favored by thistles and by wasps and bees. Here, without fear of being troubled by the passersby, I could consult the Ammophila and the Sphex [two digger or hunting wasps] and engage in that difficult conversation whose questions and answers have experiment for their language; here, without distant expeditions that take up my time, without tiring rambles that strain my nerves, I could contrive my plans of attack, lay my ambushes and watch their effects at every hour of the day. Hoc erat in votis. Yes, this was my wish, my dream, always cherished, always vanishing into the mists of the future.

And it is no easy matter to acquire a laboratory in the open fields, when harassed by a terrible anxiety about one's daily bread. For forty years have I fought, with steadfast courage, against the paltry plagues of life; and the long-wished-for laboratory has come at last. What it has cost me in perseverance and relentless work I will not try to say. It has come; and, with it—a more serious condition—perhaps a little leisure. I say perhaps, for my leg is still hampered with a few links of the convict's chain.

The wish is realized. It is a little late, O my pretty insects! I greatly fear that the peach is offered to me when I am beginning to have no teeth wherewith to eat it. Yes, it is a little late: the wide horizons of the outset have shrunk into a low and stifling canopy, more and more straitened day by day. Regretting nothing in the past, save those whom I have lost; regretting nothing, not even my first youth; hoping nothing either, I have reached the point at which, worn out by the experience of things, we ask ourselves if life be worth the living.

Amid the ruins that surround me, one strip of wall remains standing, immovable upon its solid base: my passion for scientific truth. Is that enough, O my busy insects, to enable me to add yet a few seemly pages to your history? Will my strength not cheat my good intentions? Why, indeed, did I forsake you so long? Friends have reproached me for it. Ah, tell them, tell those friends, who are yours as well as mine, tell them that it was not forgetfulness on my part, not weariness, nor neglect: I thought of you; I was convinced that the Cerceris [a digger wasp] cave had more fair secrets to reveal to us, that the chase of the Sphex held fresh surprises in store. But time failed me; I was alone, deserted, struggling against misfortune. Before philosophizing, one had to live. Tell them that; and they will pardon me.

Others again have reproached me with my style, which has not the solemnity, nay, better, the dryness of the schools. They fear lest a page that is read without fatigue should not always be the expression of the truth. Were I to take their word for it, we are profound only on condition of being obscure. Come here, one and all of you—you, the sting bearers, and you, the wing-cased armor-clads—take up my defense and bear witness in my favor. Tell of the intimate terms on which I live with you, of the patience with which I observe you, of the care with which I record your actions. Your evidence is unanimous: yes, my pages, though they bristle not with hollow formulas nor learned smatterings, are the exact narrative of facts observed, neither more nor less; and whoever cares to question you in his turn will, obtain the same replies.

And then, my dear insects, if you cannot convince those good people, because you do not carry the weight of tedium, I, in my turn, will say to them: 'You rip up the animal and I study it alive; you turn it into an object of horror and pity, whereas I cause it to be loved; you labor in a torture chamber and dissecting room, I make my observations under the blue sky to the song of the cicadas, you subject cell and protoplasm to chemical tests, I study instinct in its loftiest manifestations; you pry into death, I pry into life. And why should I not complete my thought: the boars have muddied the clear stream; natural history, youth's glorious study, has, by dint of cellular improvements, become a hateful and repulsive thing. Well, if I write for men of learning, for philosophers, who, one day, will try to some extent to unravel the tough problem of instinct, I write also, I write above all things for the young. I want to make them love the natural history which you make them hate; and that is why, while keeping strictly to the domain of truth, I avoid your scientific prose, which too often, alas seems borrowed from some Iroquois idiom.

But this is not my business for the moment: I want to speak of the bit of land long cherished in my plans to form a laboratory of living entomology, the bit of land which I have at last obtained in the solitude of a little village. It is a harmas, the name given, in this district [the country round Serignan, in Provence], to an untilled, pebbly expanse abandoned to the vegetation of the thyme. It is too poor to repay the work of the plow; but the sheep passes there in spring, when it has chanced to rain and a little grass shoots up.

My harmas, however, because of its modicum of red earth swamped by a huge mass of stones, has received a rough first attempt at cultivation: I am told that vines once grew here. And, in fact, when we dig the ground before planting a few trees, we turn up, here and there, remains of the precious stock, half carbonized by time. The three pronged fork, therefore, the only implement of husbandry that can penetrate such a soil as this, has entered here; and I am sorry, for the primitive vegetation has disappeared. No more thyme, no more lavender, no more clumps of kermes oak, the dwarf oak that forms forests across which we step by lengthening our stride a little. As these plants, especially the first two, might be of use to me by offering the Bees and Wasps a spoil to forage, I am compelled to reinstate them in the ground whence they were driven by the fork.

What abounds without my mediation is the invaders of any soil that is first dug up and then left for a long time to its own resources. We have, in the first rank, the couch grass, that execrable weed which three years of stubborn warfare have not succeeded in exterminating. Next, in respect of number, come the centauries, grim looking one and all, bristling with prickles or starry halberds. They are the yellow-flowered centaury, the mountain centaury, the star thistle and the rough centaury: the first predominates. Here and there, amid their inextricable confusion, stands, like a chandelier with spreading, orange flowers for lights, the fierce Spanish oyster plant, whose spikes are strong as nails. Above it, towers the Illyrian cotton thistle, whose straight and solitary stalk soars to a height of three to six feet and ends in large pink tufts. Its armor hardly yields before that of the oyster plant. Nor must we forget the lesser thistle tribe, with first of all, the prickly or 'cruel' thistle, which is so well armed that the plant collector knows not where to grasp it; next, the spear thistle, with its ample foliage, ending each of its veins with a spear head; lastly, the black knapweed, which gathers itself into a spiky knot. In among these, in long lines armed with hooks, the shoots of the blue dewberry creep along the ground. To visit the prickly thicket when the Wasp goes foraging, you must wear boots that come to mid-leg or else resign yourself to a smarting in the calves. As long as the ground retains a few remnants of the vernal rains, this rude vegetation does not lack a certain charm, when the pyramids of the oyster plant and the slender branches of the cotton thistle rise above the wide carpet formed by the yellow-flowered centaury saffron heads; but let the droughts of summer come and we see but a desolate waste, which the flame of a match would set ablaze from one end to the other. Such is, or rather was, when I took possession of it, the Eden of bliss where I mean to live henceforth alone with the insect. Forty years of desperate struggle have won it for me.

Eden, I said; and, from the point of view that interests me, the expression is not out of place. This cursed ground, which no one would have had at a gift to sow with a pinch of turnip seed, is an earthly paradise for the bees and wasps. Its mighty growth of thistles and centauries draws them all to me from everywhere around. Never, in my insect hunting memories, have I seen so large a population at a single spot; all the trades have made it their rallying point. Here come hunters of every kind of game, builders in clay, weavers of cotton goods, collectors of pieces cut from a leaf or the petals of a flower, architects in pasteboard, plasterers mixing mortar, carpenters boring wood, miners digging underground galleries, workers handling goldbeater's skin and many more.

Who is this one? An Anthidium [a tailor bee]. She scrapes the cobwebby stalk of the yellow-flowered centaury and gathers a ball of wadding which she carries off proudly in the tips of her mandibles. She will turn it, under ground, into cotton felt satchels to hold the store of honey and the egg. And these others, so eager for plunder? They are Megachiles [leaf-cutting bees], carrying under their bellies their black, white or blood red reaping brushes. They will leave the thistles to visit the neighboring shrubs and there cut from the leaves oval pieces which will be made into a fit receptacle to contain the harvest. And these, clad in black velvet? They are Chalicodomae [mason bees], who work with cement and gravel. We could easily find their masonry on the stones in the harmas. And these noisily buzzing with a sudden flight? They are the Anthophorae [wild bees], who live in the old walls and the sunny banks of the neighborhood.

Now come the Osmiae. One stacks her cells in the spiral staircase of an empty snail shell; another, attacking the pith of a dry bit of bramble, obtains for her grubs a cylindrical lodging and divides it into floors by means of partition walls; a third employs the natural channel of a cut reed; a fourth is a rent-free tenant of the vacant galleries of some mason bee. Here are the Macrocerae and the Eucerae, whose males are proudly horned; the Dasypodae, who carry an ample brush of bristles on their hind legs for a reaping implement; the Andrenae, so manifold in species; the slender-bellied Halicti [all wild bees]. I omit a host of others. If I tried to continue this record of the guests of my thistles, it would muster almost the whole of the honey yielding tribe. A learned entomologist of Bordeaux, Professor Perez, to whom I submit the naming of my prizes, once asked me if I had any special means of hunting, to send him so many rarities and even novelties. I am not at all an experienced and, still less, a zealous hunter, for the insect interests me much more when engaged in its work than when struck on a pin in a cabinet. The whole secret of my hunting is reduced to my dense nursery of thistles and centauries.

By a most fortunate chance, with this populous family of honey gatherers was allied the whole hunting tribe. The builders' men had distributed here and there in the harmas great mounds of sand and heaps of stones, with a view to running up some surrounding walls. The work dragged on slowly; and the materials found occupants from the first year. The mason bees had chosen the interstices between the stones as a dormitory where to pass the night, in serried groups. The powerful eyed lizard, who, when close pressed, attacks both man and dog, wide mouthed, had selected a cave wherein to lie in wait for the passing scarab [a dung beetle also known as the sacred beetle]; the black-eared chat, garbed like a Dominican, white-frocked with black wings, sat on the top stone, singing his short rustic lay: his nest, with its sky blue eggs, must be somewhere in the heap. The little Dominican disappeared with the loads of stones. I regret him: he would have been a charming neighbor. The eyed lizard I do not regret at all.

The sand sheltered a different colony. Here, the Bembeces [digger wasps] were sweeping the threshold of their burrows, flinging a curve of dust behind them; the Languedocian Sphex was dragging her Ephippigera [a green grasshopper] by the antennae; a Stizus [a hunting wasp] was storing her preserves of Cicadellae [froghoppers]. To my sorrow, the masons ended by evicting the sporting tribe; but, should I ever wish to recall it, I have but to renew the mounds of sand: they will soon all be there.

Hunters that have not disappeared, their homes being different, are the Ammophilae, whom I see fluttering, one in spring, the others in autumn, along the garden walks and over the lawns, in search of a caterpillar; the Pompili [digger or hunting wasp], who travel alertly, beating their wings and rummaging in every corner in quest of a spider. The largest of them waylays the Narbonne Lycosa [known also as the black-bellied tarantula], whose burrow is not infrequent in the harmas. This burrow is a vertical well, with a curb of fescue grass intertwined with silk. You can see the eyes of the mighty Spider gleam at the bottom of the den like little diamonds, an object of terror to most. What a prey and what dangerous hunting for the Pompilus! And here, on a hot summer afternoon, is the Amazon ant, who leaves her barrack rooms in long battalions and marches far afield to hunt for slaves. We will follow her in her raids when we find time. Here again, around a heap of grasses turned to mould, are Scoliae [large hunting wasps] an inch and a half long, who fly gracefully and dive into the heap, attracted by a rich prey, the grubs of Lamellicorns, Orycotes and Ceotoniae [various beetles].

What subjects for study! And there are more to come. The house was as utterly deserted as the ground. When man was gone and peace assured, the animal hastily seized on everything. The warbler took up his abode in the lilac shrubs; the greenfinch settled in the thick shelter of the cypresses; the sparrow carted rags and straw under every slate; the Serin finch, whose downy nest is no bigger than half an apricot, came and chirped in the plane tree tops; the Scops made a habit of uttering his monotonous, piping note here, of an evening; the bird of Pallas Athene, the owl, came hurrying along to hoot and hiss.

In front of the house is a large pond, fed by the aqueduct that supplies the village pumps with water. Here, from half a mile and more around, come the frogs and Toads in the lovers' season. The natterjack, sometimes as large as a plate, with a narrow stripe of yellow down his back, makes his appointments here to take his bath; when the evening twilight falls, we see hopping along the edge the midwife toad, the male, who carries a cluster of eggs, the size of peppercorns, wrapped round his hindlegs: the genial paterfamilias has brought his precious packet from afar, to leave it in the water and afterwards retire under some flat stone, whence he will emit a sound like a tinkling bell. Lastly, when not croaking amid the foliage, the tree frogs indulge in the most graceful dives. And so, in May, as soon as it is dark, the pond becomes a deafening orchestra: it is impossible to talk at table, impossible to sleep. We had to remedy this by means perhaps a little too rigorous. What could we do? He who tries to sleep and cannot needs becomes ruthless.

Bolder still, the wasp has taken possession of the dwelling house. On my door sill, in a soil of rubbish, nestles the white-banded Sphex: when I go indoors, I must be careful not to damage her burrows, not to tread upon the miner absorbed in her work. It is quite a quarter of a century since I last saw the saucy cricket hunter. When I made her acquaintance, I used to visit her at a few miles' distance: each time, it meant an expedition under the blazing August sun. Today, I find her at my door; we are intimate neighbors. The embrasure of the closed window provides an apartment of a mild temperature for the Pelopaeus [a mason wasp]. The earth-built nest is fixed against the freestone wall. To enter her home, the spider huntress uses a little hole left open by accident in the shutters. On the moldings of the Venetian blinds, a few stray mason bees build their group of cells; inside the outer shutters, left ajar, a Eumenes [a mason wasp] constructs her little earthen dome, surmounted by a short, bell-mouthed neck. The common wasp and the Polistes [a solitary wasp] are my dinner guests: they visit my table to see if the grapes served are as ripe as they look.

Here, surely—and the list is far from complete—is a company both numerous and select, whose conversation will not fail to charm my solitude, if I succeed in drawing it out. My dear beasts of former days, my old friends, and others, more recent acquaintances, all are here, hunting, foraging, building in close proximity. Besides, should we wish to vary the scene of observation, the mountain [Ventoux] is but a few hundred steps away, with its tangle of arbutus, rock roses and arborescent heather; with its sandy spaces dear to the Bembeces; with its marly slopes exploited by different wasps and bees. And that is why, foreseeing these riches, I have abandoned the town for the village and come to Serignan to weed my turnips and water my lettuces.

Laboratories are being founded, at great expense, on our Atlantic and Mediterranean coasts, where people cut up small sea animals, of but meager interest to us; they spend a fortune on powerful microscopes, delicate dissecting instruments, engines of capture, boats, fishing crews, aquariums, to find out how the yolk of an Annelid's egg is constructed, a question whereof I have never yet been able to grasp the full importance; and they scorn the little land animal, which lives in constant touch with us, which provides universal psychology with documents of inestimable value, which too often threatens the public wealth by destroying our crops. When shall we have an entomological laboratory for the study not of the dead insect, steeped in alcohol, but of the living insect; a laboratory having for its object the instinct, the habits, the manner of living, the work, the struggles, the propagation of that little world, with which agriculture and philosophy have most seriously to reckon?

To know thoroughly the history of the destroyer of our vines might perhaps be more important than to know how this or that nerve fiber of a Cirriped [sea animals with hair-like legs, including the barnacles and acorn shells] ends; to establish by experiment the line of demarcation between intellect and instinct; to prove, by comparing facts in the zoological progression, whether human reason be an irreducible faculty or not: all this ought surely to take precedence of the number of joints in a Crustacean's antenna. These enormous questions would need an army of workers; and we have not one. The fashion is all for the Mollusk and the Zoophytes [plant-like sea animals, including starfishes, jellyfishes, sea anemones and sponges]. The depths of the sea are explored with many drag nets; the soil which we tread is consistently disregarded. While waiting for the fashion to change, I open my harmas laboratory of living entomology; and this laboratory shall not cost the ratepayers one farthing.

CHAPTER II. THE ANTHRAX

I made the acquaintance of the Anthrax in 1855 at Carpentras, at the time when the life history of the oil beetles was causing me to search the tall slopes beloved of the Anthophora bees [mason bees]. Her curious pupae, so powerfully equipped to force an outlet for the perfect insect incapable of the least effort, those pupae armed with a multiple plowshare at the fore, a trident at the rear and rows of harpoons on the back wherewith to rip open the Osmia bee's cocoon and break through the hard crust of the hillside, betokened a field that was worth cultivating. The little that I said about her at the time brought me urgent entreaties: I was asked for a circumstantial chapter on the strange fly. The stern necessities of life postponed to an ever retreating future my beloved investigations, so miserably stifled. Thirty years have passed; at last, a little leisure is at hand; and here, in the harmas of my village, with an ardor that has in no wise grown old, I have resumed my plans of yore, still alive like the coal smoldering under the ashes. The Anthrax has told me her secrets, which I in my turn am going to divulge. Would that I could address all those who cheered me on this path, including first and foremost the revered Master of the Landes [Leon Dufour]. But the ranks have thinned, many have been promoted to another world and their disciple lagging behind them can but record, in memory of those who are no more, the story of the insect clad in deepest mourning.

In the course of July, let us give a few sideward knocks to the bracing pebbles and detach the nests of the Chalicodoma of the Walls [a mason bee] from their supports. Loosened by the shock, the dome comes off cleanly, all in one piece. Moreover—and this is a great advantage—the cells come into view wide open on the base of the exposed nest, for at this point they have no other wall than the surface of the pebble. In this way, without any scraping, which would be wearisome work for the operator and dangerous to the inhabitants of the dome, we have all the cells before our eyes, together with their contents, consisting of a silky, amber-yellow cocoon, as delicate and translucent as an onion peeling. Let us split the dainty wrapper with the scissors, chamber by chamber, nest by nest. If fortune be at all propitious, as it always is to the persevering, we shall end by finding that the cocoons harbor two larvae together, one more or less faded in appearance, the other fresh and plump. We shall also find some, no less plentiful, in which the withered larva is accompanied by a family of little grubs wriggling uneasily around it.

Examination at once reveals the tragedy that is happening under the cover of the cocoon. The flacid and faded larva is the mason bee's. A month ago, in June, having finished its mess of honey, it wove its silken sheath for a bedchamber wherein to take the long sleep which is the prelude to the metamorphosis. Bulging with fat, it is a rich and defenseless morsel for whoever is able to reach it. Then, in spite of apparently insurmountable obstacles, the mortar wall and the tent without an opening, the flesh-eating larvae appeared in the secret retreat and are now glutting themselves on the sleeper. Three different species take part in the carnage, often in the same nest, in adjoining cells. The diversity of shapes informs us of the presence of more than one enemy; the final stage of the creatures will tell us the names and qualities of the three invaders.

Forestalling the secrets of the future for the sake of greater clearness, I will anticipate the actual facts and come at once to the results produced. When it is by itself on the body of the mason bee's larva, the murderous grub belongs either to Anthrax trifasciata, MEIGEN, or to Leucospis gigas, FAB. But, if numerous little worms, often a score and more, swarm around the victim, then it is a Chalcidid's family which we have before us. Each of these ravagers shall have its biography. Let us begin with the Anthrax.

And first the grub, as it is after consuming its victim, when it remains the sole occupant of the mason bee's cocoon. It is a naked worm, smooth, legless and blind, of a creamy dead white, each segment a perfect ring, very much curved when at rest, but with the tendency to become almost straight when disturbed. Through the diaphanous skin, the lens distinguishes patches of fat, which are the cause of its characteristic coloring. When younger, as a tiny grub a few millimeters long, it is streaked with two different kinds of stains, some white, opaque and of a creamy tint, others translucent and of the palest amber. The former come from adipose masses in course of formation; the second from the nourishing fluid or from the blood which laves those masses.

Including the head, I count thirteen segments. In the middle of the body these segments are well marked, being separated by a slight groove; but in the forepart they are difficult to count. The head is small and is soft, like the rest of the body, with no sign of any mouth parts even under the close scrutiny of the lens. It is a white globule, the size of a tiny pin's head and continued at the back by a pad a little larger, from which it is separated by a scarcely appreciable crease. The whole is a sort of nipple swelling slightly on the upper surface; and its double structure is so difficult to perceive that at first we take it for the animal's head alone, though it includes both the head and the prothorax, or first segment of the thorax.

The mesothorax, or middle segment of the thorax, which is two or three times larger in diameter, is flattened in front and separated from the nipple formed by the prothorax and the head by a deep, narrow, curved fissure. On its front surface are two pale red stigmata, or respiratory orifices, placed pretty close together. The metathorax, or last segment of the thorax, is a little larger still in diameter and protrudes. These abrupt increases in circumference result in a marked hump, sloping sharply towards the front. The nipple of which the head forms part is set at the bottom of this hump.

After the metathorax, the shape becomes regular and cylindrical, while decreasing slightly in girth in the last two or three segments. Close to the line of separation of the last two rings, I am able to distinguish, not without difficulty, two very small stigmata, just a little darker in color. They belong to the last segment. In all, four respiratory orifices, two in front and two behind, as is the rule among Flies. The length of the full sized larva is 15 to 20 millimeters and its breadth 5 to 6.

Remarkable in the first place by the protuberance of its thorax and the smallness of its head, the grub of the Anthrax acquires exceptional interest by its manner of feeding. Let us begin by observing that, deprived of all, even the most rudimentary walking apparatus, the animal is absolutely incapable of shifting its position. If I disturb its rest, it curves and straightens itself in turns by a series of contractions, it tosses about violently where it lies, but does not manage to progress. It fidgets and gets no farther. We shall see later the magnificent problem raised by this inertness.

For the moment, a most unexpected fact claims all our attention. I refer to the extreme readiness with which the Anthrax' larva quits and returns to the Chalicodoma grub on which it is feeding. After witnessing flesh eating larvae at hundreds and hundreds of meals, I suddenly find myself confronted with a manner of eating that bears no relation to anything which I have seen before. I feel myself in a world that baffles my old experience. Let us recall the table manners of a larva living on prey, the Ammophila's for instance, when devouring its caterpillar. A hole is made in the victim's side; and the head and neck of the nursling dive deep into the wound, to root luxuriously among the entrails. There is never a withdrawal from the gnawed belly, never a recoil to interrupt the feast and to take breath awhile. The vivacious animal always goes forward, chewing, swallowing, digesting, until the caterpillar's skin is emptied of its contents. Once seated at table, it does not budge as long as the victuals last. To tease it with a straw is not always enough to induce it to withdraw its head outside the wound; I have to use violence. When removed by force and then left to its own devices, the creature hesitates for a long time, stretches itself and mouths around, without trying to open a passage through a new wound. It needs the attacking point that has just been abandoned. If it finds the spot, it makes its way in and resumes the work of eating; but its future is jeopardized from this time forward, for the game, now perhaps tackled at inopportune points, is liable to go bad.

With the Anthrax' grub, there is none of this mangling, none of this persistent clinging to the entrance wound. I have but to tease it with the tip of a hair pencil and forthwith it retires; and the lens reveals no wound at the abandoned spot, no such effusion of blood as there would be if the skin were perforated. When its sense of security is restored, the grub once more applies its pimple head to the fostering larva, at any point, no matter where; and, so long as my curiosity does not prevent it, keeps itself fixed there, without the least effort, or the least perceptible movement that could account for the adhesion. If I repeat the touch with the pencil, I see the same sudden retreat and, soon after, the same contact just as readily renewed.

This facility for gripping, quitting and regripping, now here, now there and always without a wound, the part of the victim whence the nourishment is drawn tells us of itself that the mouth of the Anthrax is not armed with mandibular fangs capable of digging into the skin and tearing it. If the flesh were gashed by any such pincers, one or two attempts would be necessary before they could be released or reapplied; besides, each point bitten would display a lesion. Well, there is nothing of the kind: a conscientious examination through the magnifying glass shows conclusively that the skin is intact; the grub glues its mouth to its prey or withdraws it with an ease that can only be explained by a process of simple contact. This being so, the Anthrax does not chew its food as do the other carnivorous grubs; it does not eat, it inhales.

This method of taking nourishment implies an exceptional apparatus of the mouth, into which it behooves us to inquire before continuing. My most powerful magnifying glass at last discovers, at the center of the pimple head, a small spot of an amber-russet color; and that is all. For a more exhaustive examination we will employ the microscope. I cut off the strange pimple with the scissors, wash it in a drop of water and place it on the object slide. The mouth now stands revealed as a round spot which, for hue and for the smallness of its size, may be compared with the front stigmata. It is a small conical crater, with sides of a pale yellowish-red and with faint, more or less concentric lines. At the bottom of this funnel is the opening of the gullet, itself tinted red in front and promptly spreading into a cone at the back. There is not the slightest trace of mandibular fangs, of jaws, of mouth parts for seizing and grinding. Everything is reduced to the bowl shaped opening, with a delicate lining of horny texture, as is shown by the amber hue and the concentric streaks. When I look for some term to designate this digestive entrance, of which so far I know no other example,

I can find only that of a sucker or cupping glass. Its attack is a mere kiss, but what a perfidious kiss!

We know the machine; now let us see the working. To facilitate observation, I shifted the newborn Anthrax grub, together with the Chalicodoma grub, its wet nurse, from the natal cell into a glass tube. I was thus able, by employing as many tubes as I wanted, to follow from start to finish, in all its most intimate details, the strange repast which I am going to describe.

The worm is fixed by its sucker to any convenient part of the nurse, plump and fat as butter. It is ready to break off its kiss suddenly, should anything disquiet it, and to resume it as easily when tranquillity is restored. No Lamb enjoys greater liberty with its mother's teat. After three or four days of this contact of the nurse and nursling, the former, at first replete and endowed with the glossy skin that is a sign of health, begins to assume a withered aspect. Her sides fall in, her fresh color fades, her skin becomes covered with little folds and gives evidence of an appreciable shrinking in this breast which, instead of milk, yields fat and blood. A week is hardly past before the progress of the exhaustion becomes startlingly rapid. The nurse is flabby and wrinkled, as though borne down by her own weight, like a very slack object. If I move her from her place, she flops and sprawls like a half-filled water bottle over the new supporting plane. But the Anthrax' kiss goes on emptying her: soon she is but a sort of shriveled lard bag, decreasing from hour to hour, from which the sucker draws a few last oily drains. At length, between the twelfth and the fifteenth day, all that remains of the larva of the mason bee is a white granule, hardly as large as a pin's head.

This granule is the water bottle drained to the last drop, is the nurse's breast emptied of all its contents. I soften the meager remnant in water; then, keeping it still immersed, I blow into it through an extremely attenuated glass tube. The skin fills out, distends and resumes the shape of the larva, without there being an outlet anywhere for the compressed air. It is intact, therefore; it is free of any perforation, which would be forthwith revealed under the water by an escape of gas. And so, under the Anthrax' cupping glass, the oily bottle has been drained by a simple transpiration through the membrane; the substance of the nurse grub has been transfused into the body of the nursling by a process akin to that known in physics as endosmosis. What should we say to a method of being suckled by the mere application of the mouth to a teatless breast? What we see here may be compared with that: without any outlet, the milk of the Chalicodoma grub passes into the stomach of the Anthrax' larva.

Is it really an instance of endosmosis? Might it not rather be atmospheric pressure that stimulates the flow of nourishing fluids and distils them into the Anthrax' cup-shaped mouth, working, in order to create a vacuum, almost like the suckers of the Cuttlefish? All this is possible, but I shall refrain from deciding, preferring to assign a large share to the unknown in this extraordinary method of nutrition. It ought, I think, to provide physiologists with a field of research in which new views on the hydrodynamics of live fluids might well be gleaned; and this field trenches upon others that would also yield rich harvests. The brief span of my days compels me to set the problem without seeking to solve it.

And the second problem is this: the Chalicodoma grub destined to feed the Anthrax is without a wound of any kind. The mother of the tiny larva is a feeble Fly deprived of whatsoever weapon capable of injuring her offspring's prey. Moreover, she is absolutely powerless to penetrate the mason bee's fortress, powerless as a fluff of down against a rock. On this point there is no doubt: the future wet nurse of the Anthrax has not been paralyzed as are the live provisions collected by the Hunting Wasps; she has received no bite nor scratch nor contusion of any sort; she has experienced nothing out of the common: in short, she is in her normal state. The billeted nursling arrives, we shall presently see how; he arrives, scarcely visible, almost defying the scrutiny of the lens; and, having made his preparations, he installs himself, he, the atom, upon the monstrous nurse, whom he is to drain to the very husk. And she, not paralyzed by a preliminary vivisection, endowed with all her normal vitality, lets him have his way, lets herself be sucked dry, with the utmost apathy. Not a tremor in her outraged flesh, not a quiver of resistance. No corpse could show greater indifference to the bite which it receives.

Ah, but the maggot has chosen the hour of attack with traitorous cunning! Had it appeared upon the scene earlier, when the larva was consuming its store of honey, things of a surety would have gone badly with it. The assaulted one, feeling herself bled to death by that ravenous kiss, would have protested with much wriggling of body and grinding of mandibles. The position would have ceased to be tenable and the intruder would have perished. But at this hour all danger has disappeared. Enclosed in its silken tent, the larva is seized with the lethargy that precedes the metamorphosis. Its condition is not death, but neither is it life. It is an intermediary condition; it is almost the latent vitality of grain or egg. Therefore there is no sign of irritation on the larva's part under the needle with which I stir it and still less under the sucker of the Anthrax grub, which is able to drain the affluent breast in perfect safety.

This lack of resistance, induced by the torpor of the transformation, appears to me necessary, in view of the weakness of the nursling as it leaves the egg, whenever the mother is herself incapable of depriving the victim of the power of self defense. And so the nonparalyzed larvae are attacked during the period of the nymphosis. We shall soon see other instances of this.

Motionless though it be, the Chalicodoma grub is none the less alive. The primrose tint and the glossy skin are unequivocal signs of health: Were it really dead, it would, in less than twenty-four hours, turn a dirty brown and, soon after, decompose into a fluid putrescence. Now here is the marvelous thing: during the fortnight, roughly, that the Anthrax' meal lasts, the butter color of the larva, an unfailing symptom of the presence of life, continues unaltered and does not change into brown, the sign of putrefaction, until hardly anything remains; and even then the brown hue is often absent. As a rule, the look of live flesh is preserved until the final pellet, formed of the skin, the sole residue, makes its appearance. This pellet is white, with not a speck of tainted matter, proving that life persists until the body is reduced to nothing.

We here witness the transfusion of one animal into another, the change of Chalicodoma substance into Anthrax substance; and, as long as the transfusion is not complete, as long as the eaten has not disappeared altogether and become the eater, the ruined organism fights against destruction. What manner of life is this, which may be compared with the life of a night light whose extinction is not accomplished until the last drop of oil has burnt away? How is any creature able to fight against the final tragedy of corruption up to the last moment in which a nucleus of matter remains as the seat of vital energy? The forces of the living creature are here dissipated not through any disturbance of the equilibrium of those forces, but for the want of any point of application for them: the larva dies because materially there is no more of it.

Can we be in the presence of the diffusive life of the plant, a life which persists in a fragment? By no means: the grub is a more delicate organic structure. There is unity between the several parts; and none of them can be jeopardized without involving the ruin of the others. If I myself give the larva a wound, if I bruise it, the whole body very soon turns brown and begins to rot. It dies and decomposes by the mere prick of a needle; it keeps alive, or at least preserves the freshness of the live tissues, so long as it is not entirely emptied by the Anthrax' sucker. A nothing kills it; an atrocious wasting does not. No, I fail to understand the problem; and I bequeath it to others.

All that I can see by way of a glimpse—and even then I put forward my suspicions with extreme reserve—all that I am permitted to surmise is reduced to this: the substance of the sleeping larva as yet has no very definite static existence; it is like the raw materials collected for a building; it is waiting for the elaboration that is to make a bee of it. To mould those shapeless lumps of the future insect, the air, that prime adjuster of living things, circulates among them, passing through a network of ducts. To organize them, to direct the placing of them, the nervous system, the embryo of the animal, distributes its ramifications over them. Nerve and air duct, therefore, are the essentials; the rest is so much material in reserve for the process of the metamorphosis. As long as that material is not employed, as long as it has not acquired its final equilibrium, it can grow less and less; and life, though languishing, will continue all the same on the express condition that the respiratory organs and the nervous filaments be respected. It is as it were the flame of the lamp, which, whether full or empty, continues to give light so long as the wick is soaked in oil. Nothing but fluids, the plastic materials held in reserve, can be distilled by the Anthrax' sucker through the unpierced skin of the grub; no part of the respiratory and nervous systems passes. As the two essential functions remain unscathed, life goes on until exhaustion is completed. On the other hand, if I myself injure the larva, I disturb the nervous or air conducting filaments; and the bruised part spreads a taint, followed by putrefaction, all over the body.

I have elsewhere, speaking of the Scolia [a digger wasp] devouring the Cetonia grub, enlarged upon this refined art of eating which consists in consuming the prey while killing it only at the last mouthfuls. The Anthrax has the same requirements as his competitors who dine off fresh viands. He needs meat of that day, taken from a single joint that has to last a fortnight without going bad. His method of consuming reaches the highest level of art: he does not cut into his prey, he sips it little by little through his sucker. In this way, any dangerous risk is averted. Whether he imbibe at this spot or at that, even if he abandon the sucking process and resume it later, by no accident can he ever attack that which it is incumbent upon him to respect lest corruption supervene. The others have a fixed position on the victim, a place at which their mandibles have to bite and enter. If they move away from it, if they miss the appointed path, they imperil their existence. The Anthrax, more highly favored, puts his mouth where it suits him; he leaves off when he pleases and when he pleases starts again.

Unless I labor under a delusion, I think that I see the necessity for this privilege. The egg of the carnivorous burrower is firmly fixed on the victim at a point which varies considerably, it is true, according to the nature of the prey, but which is uniform for the same species of prey; moreover—and this is an important condition—the point of adhesion of that egg is always the head, whereas the egg of a bee, of the Osmia, for instance, is fixed to the mess of honey by the hinder end. When hatched, the new born

Wasp grub has not to choose for itself, at its risk and peril, the suitable point at which to take the first cut in the quarry without fear of killing it too quickly: all that it need do is to bite at the spot where it has just been born. The mother, with her unfailing instinct, has already made the dangerous choice; she has stuck her egg on the propitious spot and, by the very act of doing so, marked out the course for the inexperienced grub to follow. The tact of ripe age here guides the young larva's behavior at table.

The conditions are very different in the Anthrax' case. The egg is not placed upon the victuals, it is not even laid in the mason bee's cell. This is the natural consequence of the mother's feeble frame and of her lack of any instrument, such as a probe or auger, capable of piercing the mortar wall. It is for the newly hatched grub to make its own way into the dwelling. It enters, finds itself in the presence of ample provisions, the larva of the mason bee. Free of its actions, it is at liberty to attack the prey where it chooses; or rather the attacking point will be decided at haphazard by the first contact of the mouth in quest of food. Grant this mouth a set of carving tools, jaws and mandibles; in short, suppose the grub of the Fly to possess a manner of eating similar to that of the other carnivorous larvae; and the nursling is at once threatened with a speedy death. He will split open his nurse's belly, he will dig without any rule to guide him, he will bite at random, essentials as well as accessories; and, from one day to the next, he will set up gangrene in the violated mass, even as I myself do when I give it a wound. For the lack of an attacking point prescribed for him at birth, he will perish on the damaged provisions. His freedom of action will have killed him.

Certainly, liberty is a noble attribute, even in an insignificant grub; but it also has its dangers everywhere. The Anthrax escapes the peril only on the condition of being, so to speak, muzzled. His mouth is not a fierce forceps that tears asunder; it is a sucker that exhausts but does not wound. Thus restrained by this safety appliance, which changes the bite into a kiss, the grub has fresh victuals until it has finished growing, although it knows nothing of the rules of methodical consumption at a fixed point and in a predetermined direction.

The considerations which I have set forth seem to me strictly logical: the Anthrax, owing to the very fact that he is free to take his nourishment where he pleases on the body of the fostering larva, must, for his own protection, be made incapable of opening his victim's body. I am so utterly convinced of this harmonious relation between the eater and the eaten that I do not hesitate to set it up as a principle. I will therefore say this: whenever the egg of any kind of insect is not fastened to the larva destined for its food, the young grub, free to select the attacking point and to change it at will, is as it were muzzled and consumes its provisions by a sort of suction, without inflicting any appreciable wound. This restriction is essential to the maintenance of the victuals in good condition. My principle is already supported by examples many and various, whose depositions are all to the same effect. The witnesses include, after the Anthrax, the Leucospis [a parasitic insect] and his rivals, whose evidence we shall hear presently; the Ephialtes mediator [an Ichneumon fly], who feeds, in the dry brambles, on the larva of the Black Psen [a digger wasp]; the Myodites, that strange, fly-shaped beetle whose grub consumes the larva of the cockchafer. All—flies, ichneumon flies and beetles—scrupulously spare their foster mother; they are careful not to tear her skin, so that the vessel may keep its liquid good to the last.

The wholesomeness of the victuals is not the only condition imposed: I find a second, which is no less essential. The substance of the fostering larva must be sufficiently fluid to ooze through the unbroken skin under the action of the sucker. Well, the necessary fluidity is realized as the time of the metamorphosis draws near. When they wished Medea to restore Pelias to the vigor of youth, his daughters cut the old king's body to pieces and boiled it in a cauldron, for there can be no new existence without a prior dissolution. We must pull down before we can rebuild; the analysis of death is the first step towards the synthesis of life. The substance of the grub that is to be transformed into a bee begins, therefore, by disintegrating and dissolving into a fluid broth. The materials of the future insect are obtained by a general recasting. Even as the founder puts his old bronzes into the melting pot in order afterwards to cast them in a mould whence the metal will issue in a different shape, so life liquefies the grub, a mere digesting machine, now thrown aside, and out of its running matter produces the perfect insect, bee, butterfly or beetle, the final manifestation of the living creature.

Let us open a Chalicodoma grub under the microscope, during the period of torpor. Its contents consists almost entirely of a liquid broth, in which swim numberless oily globules and a fine dust of uric acid, a sort of off-throw of the oxidized tissues. A flowing thing, shapeless and nameless, is all that the animal is, if we add abundant ramified air ducts, some nervous filaments and, under the skin, a thin layer of muscular fibers. A condition of this kind accounts for a fatty transpiration through the skin when the Anthrax' sucker is at work. At any other time, when the larva is in the active period or else when the insect has reached the perfect stage, the firmness of the tissues would resist the transfusion and the suckling of the Anthrax would become a difficult matter, or even impossible. In point of fact, I find the grub of the fly established, in the vast majority of cases, on the sleeping larva and sometimes, but rarely, on the pupa. Never do I see it on the vigorous larva eating its honey; and hardly ever on the insect brought to perfection, as we find it enclosed in its cell all through the autumn and winter. And we can say the same of the other grub eaters that drain their victims without wounding them: all are engaged in their death dealing work during the period of torpor, when the tissues are fluidified. They empty their patient, who has become a bag of running grease with a diffused life; but not one, among those I know, reaches the Anthrax' perfection in the art of extraction.

Nor can any be compared with the Anthrax as regards the means brought into play in order to leave the cell. These others, when they become perfect insects, have implements for sapping and demolishing, stout mandibles, capable of digging the ground, of pulling down clay partition walls and even of reducing the mason bee's tough cement to powder. The Anthrax, in her final form, has nothing like this. Her mouth is a short, soft proboscis, good at most for soberly licking the sugary exudations of the flowers; her slim legs are so feeble that to move a grain of sand were an excessive task for them, enough to strain every joint; her great, stiff wings, which must remain full spread, do not allow her to slip through a narrow passage; her delicate suit of downy velvet, from which you take the bloom by merely breathing on it, could not withstand the rough contact of the gallery of a mine. Unable herself to enter the Mason bee's cell to lay her egg, she cannot leave it either, when the time comes to free herself and appear in broad daylight in her wedding dress. The larva, on its side, is powerless to prepare the way for the coming flight. That buttery little cylinder, owning no tools but a sucker so flimsy that it barely arrives at substance and so small that it is almost a geometrical point, is even weaker than the adult insect, which at least flies and walks. The Mason bee's cell represents to it a granite cave. How to get out? The problem would be insoluble to those two incapables, if nothing else played its part.

Among insects, the nymph, or pupa, the transition stage between the larval and the adult form, is generally a striking picture of every weakness of a budding organism. A sort of mummy tight bound in swaddling clothes, motionless and impassive, it awaits the resurrection. Its tender tissues flow in every direction; its limbs, transparent as crystal, are held fixed in their place, along the side, lest a movement should disturb the exquisite delicacy of the work in course of accomplishment. Even so, to secure his recovery, is a broken boned patient held captive in the surgeon's bandages. Absolute stillness is necessary in both cases, lest they be crippled or even die.

Well, here, by a strange inversion that confuses all our views on life, a Cyclopean task is laid upon the nymph of the Anthrax. It is the nymph that has to toil, to strive, to exhaust itself in efforts to burst the wall and open the way out. To the embryo falls the desperate duty, which shows no mercy to the nascent flesh; to the adult insect the joy of resting in the sun. This transposition of functions has as its result a well sinker's equipment in the nymph, an eccentric, complicated equipment which nothing suggested in the larva and which nothing recalls in the perfect insect. The set of tools includes an assortment of plowshares, gimlets, hooks and spears and of other implements that are not found in our trades nor named in our dictionaries. Let us do our best to describe the strange piercing gear.

In a fortnight at most, the Anthrax has consumed the Chalicodoma grub, whereof naught remains but the skin, gathered into a white granule. By the time that July is nearly over, it becomes rare to find any nurslings left upon their nurses. From this period until the following May, nothing fresh happens. The Anthrax retains its larval shape without any appreciable change and lies motionless in the mason bee's cocoon, beside the pellet remains. When the fine days of May arrive, the grub shrivels and casts its skin and the nymph appears, fully clad in a stout, reddish, horny hide.

The head is round and large, separated from the thorax by a strangulated furrow, crowned on top and in front with a sort of diadem of six hard, sharp, black spikes, arranged in a semicircle whose concave side faces downward. These spikes decrease slightly in length from the summit to the ends of the arch. Taken together, they suggest the radial crowns which we see the Roman emperors of the Decadence wear on the medals. This six-fold plowshare is the chief excavating tool. Lower down, on the median line, the instrument is finished off with a separate group of two small black spikes, placed close together.

The thorax is smooth, the wing cases large, folded under the body like a scarf and coming almost to the middle of the abdomen. This has nine segments, of which four, starting with the second, are armed, on the back, down the middle, with a belt of little horny arches, pale brown in color, drawn up parallel to one another, set in the skin by their convex surfaces and finishing at both ends with a hard, black point. Altogether, the belt thus forms a double row of little thorns, with a hollow in between. I count about twenty-five twin-toothed arches to one segment, which gives a total of two hundred spikes for the four rings thus armed.

The use of this rasp, or grater, is obvious: it gives the nymph a purchase on the wall of its gallery as the work proceeds. Thus anchored on a host of points, the stern pioneer is able to hit the obstacle harder with its diadem of awls. Moreover, to make it more difficult for the instrument to recoil, long, stiff bristles, pointing backwards, are scattered here and there among the climbing belts. There are some besides on the other segments, both on the ventral and the dorsal surface. On the flanks, they are thicker and arranged as it were in clusters.

The sixth segment carries a similar belt, but a much less powerful one, consisting of a single row of unassuming thorns. The belt is weaker still on the seventh segment; lastly, on the eighth, it is reduced to a mere rough brown shading. Commencing with the sixth, the rings decrease in width and the abdomen ends in a cone, the extremity of which, formed of the ninth segment, constitutes a weapon of a new kind. It is a sheaf of eight brown spikes. The last two exceed the others in length and stand out from the group in a double terminal plowshare.

There is a round air hole in front, on either side of the thorax, and similar stigmata on the flanks of each of the first seven abdominal segments. When at rest, the nymph is curved into a bow. When about to act, it suddenly unbends and straightens itself. It measures 15 to 20 millimeters long and 4 to 5 millimeters across.

Such is the strange perforating machine that is to prepare an outlet for the feeble Anthrax through the Mason bee's cement. The structural details, so difficult to explain in words, may be summed up as follows: in front, on the forehead, a diadem of spikes, the ramming and digging tool; behind, a many bladed plowshare which fits into a socket and allows the pupa to slacken suddenly in readiness for an attack on the barrier which has to be demolished; on the back, four climbing belts, or graters, which keep the animal in position by biting on the walls of the tunnel with their hundreds of teeth; and, all over the body, long, stiff bristles, pointing backwards, to prevent falls or recoils.

A similar structure exists in the other species of Anthrax with slight variations of detail. I will confine myself to one instance, that of Anthrax sinuata, who thrives at the cost of Osmia tricornis. Her nymph differs from that of Anthrax trifasciata, the Anthrax of the mason bee, in possessing less powerful armor. Its four climbing belts consist of only fifteen to seventeen double spiked arches, instead of twenty-five; also, the abdominal segments, from the sixth onwards, are supplied merely with stiff bristles, without a trace of horny spikes. If the evolution of the various Anthrax flies were better known to us, the number of these arches would, I believe, be of great service to entomology in the differentiation of species. I see it remaining constant for any given species, with marked variations between one species and another. But this is not my business: I merely call the attention of the classifiers to this field of study and pass on.

About the end of May, the coloring of the nymph, hitherto a light red, alters greatly and forecasts the coming transformation. The head, the thorax and the scarf formed by the wings become a handsome, shiny black. A dark band shows on the back of the four segments with their two rows of spikes; three spots appear on the two next rings; the anal armor becomes darker. In this manner we foresee the black livery of the coming insect. The time has arrived for the pupa to work at the exit gallery.

I was anxious to see it in action, not under natural conditions, which would be impracticable, but in a glass tube in which I confine it between two thick stoppers of sorghum pith. The space thus marked off is about the same size as the natal cell. The partitions front and back, although not so stout as the Chalicodoma's masonry, are nevertheless firm enough not to yield except to prolonged efforts; on the other hand, the side walls are smooth and the toothed belts will not be able to grip them: a most unfavorable condition for the worker. No matter: in the space of a single day, the pupa pierces the front partition, three quarters of an inch thick. I see it fixing its double plowshare against the back partition, arching into a bow and then suddenly releasing itself and striking the plug in front of it with its barbed forehead. Under the impact of the spikes, the sorghum slowly crumbles to pieces. It is slow in coming away; but it comes away all the same, atom by atom. At long intervals, the method changes. With its crown of awls driven into the pith, the animal frets and fidgets, sways on the pivot of its anal armor. The work of the auger follows that of the pickaxe. Then the blows recommence, interspersed with periods of rest to recover from the fatigue. At last, the hole is made. The pupa slips into it, but does not pass through entirely: the head and thorax appear outside; the abdomen remains held in the gallery.

The glass cell, with its lack of supports at the side, has certainly perplexed my subject, which does not seem to have made use of all its methods. The hole through the sorghum is wide and irregular; it is a clumsy breach and not a gallery. When made through the mason bee's walls, it is cylindrical, fairly neat and exactly of the animal's diameter. So I hope that, under natural conditions, the pupa does not give quite so many blows with the pickaxe and prefers to work with the drill.

Narrowness and evenness in the exit tunnel are necessary to it. It always remains half caught in it and even pretty securely fixed by the graters on its back. Only the head and thorax emerge into the outer air. This is a last precaution for the final deliverance. A fixed support is, in fact, indispensable to the Anthrax for issuing from her horny sheath, unfurling her great wings and extricating her slender legs from their scabbards. All this very delicate work would be endangered by any lack of steadiness.

The pupa, therefore, remains fixed by the graters of its back in the narrow exit gallery and thus supplies the stable equilibrium essential to the new birth. All is ready. It is time now for the great act. A transversal cleft makes its appearance on the forehead, at the bottom of the perforating diadem; a second, but longitudinal slit divides the skull in two and extends down the thorax. Through this cross-shaped opening, the Anthrax suddenly appears, all moist with the humors of life's laboratory. She steadies herself upon her trembling legs, dries her wings and takes to flight, leaving at the window of the cell her nymphal slough, which keeps intact for a very long period. The sand-colored fly has five or six weeks before her, wherein to explore the clay nests amid the thyme and to take her small share of the joys of life. In July, we shall see her once more, busy this time with the entrance into the cell, which is even stranger than the exit.

CHAPTER III. ANOTHER PROBER (PERFORATOR)

What can he be called, this creature whose style and title I dare not inscribe at the head of the chapter? His name is Monodontomerus cupreus, SM. Just try it, for fun: Mo-no-don-to-me-rus. What a gorgeous mouthful! What an idea it gives one of some beast of the Apocalypse! We think, when we pronounce the word, of the prehistoric monsters: the mastodon, the mammoth, the ponderous megatherium. Well, we are misled by the scientific label: we have to do with a very paltry insect, smaller than the common gnat.

There are good people like that, only too happy to serve science with resounding appellations that might come from Timbuktu; they cannot name you a midge without striking terror into you. O ye wise and revered ones, ye christeners of animals, I am willing, in my study, to make use—but not undue use—of your harsh terminology, with its conglomeration of syllables; but there is a danger of their leaving the sanctum and appearing before the public, which is always ready to show its lack of deference for terms that do not respect its ears. I, wishing to speak like everybody else, so that I may be understood by all, and persuaded that science has no need of this Brobdignagian jargon, make a point of avoiding technical nomenclature when it becomes too barbarous, when it threatens to lumber the page the moment my pen attempts it. And so I abandon Monodontomerus.

It is a puny little insect, almost as tiny as the midges whom we see eddying in a ray of sunshine at the end of autumn. Its dress is golden bronze; its eyes are coral red. It carries a naked sword, that is to say, the sheath of its drill stands out slantwise at the tip of its belly, instead of lying in a hollow groove along the back, as it does with the Leucospis. This scabbard holds the latter half of the inoculating filament, which extends below the animal to the base of the abdomen. In short, its utensil is that of the Leucospis, with this difference, that its lower half sticks out like a rapier.

This mite that bears a sword upon her rump is yet another persecutor of the mason bees and not one of the least formidable. She exploits their nests at the same time as the Leucospis. I see her, like the Leucospis, slowly explore the ground with her antennae; I see her, like the Leucospis, bravely drive her dagger into the stone wall. More taken up with her work, less conscious perhaps of danger, she pays no heed to the man who is observing her so closely. Where the Leucospis flies, she does not budge. So great is her assurance that she comes right into my study, to my work table, and disputes my ownership of the nests whose occupants I am examining. She operates under my lens, she operates just beside my forceps. What risk does she run? What can one do to a thing so very small? She is so certain of her safety that I can take the Mason's nest in my hand, move it, put it down and take it up again without the insect's raising any objection: it continues its work even when my magnifying glass is placed over it.

One of these heroines has come to inspect a nest of the Chalicodoma of the Walls, most of whose cells are occupied by the numerous cocoons of a parasite, the Stelis. The contents of these cells, which have been partially ripped up to satisfy my curiosity, are very much exposed to view. The windfall appears to be appreciated, for I see the dwarf ferret about from cell to cell for four days on end, see her choose her cocoon and insert her awl in the most approved fashion. I thus learn that sight, although an indispensable guide in searching, does not decide upon the proper spot for the operation. Here is an insect exploring not the stony exterior of the mason's dwelling, but the surface of cocoons woven of silk. The explorer has never found herself placed in such circumstances, nor has any of her race before her, every cocoon, under normal conditions, being protected by a surrounding wall. No matter: despite the profound difference in the surfaces, the insect does not waver. Warned by a special sense, an undecipherable riddle to ourselves, it knows that the object of its search lies hidden under this unfamiliar casing. The sense of smell has already been shown to be out of the question; that of sight is now eliminated in its turn.

That she should bore through the cocoons of the Stelis, a parasite of the mason bee, does not surprise me at all: I know how indifferent my bold visitor is to the nature of the victuals destined for her family. I have noticed her presence in the homes of bees differing greatly in size and habits: Anthophorae, Osmiae, Chalicodomae, Anthidia. The Stelis exploited on my table is one victim more; and that is all. The interest does not lie there. The interest lies in the maneuvers of the insect, which I am able to follow under the most favorable conditions.

Bent sharply at right angles, like a couple of broken matches, the antennae feel the cocoon with their tips alone. The terminal joint is the home of this strange sense which discerns from afar what no eye sees, no scent distinguishes and no ear hears. If the point explored be found suitable, the insect hoists itself on tiptoe so as to give full scope to the play of its mechanism; it brings the tip of the belly a little forward; and the entire ovipositor—inoculating-needle and scabbard—stands perpendicular to the cocoon, in the center of the quadrilateral described by the four hind legs, an eminently favorable position for obtaining the maximum effect. For some time, the whole of the awl bears on the cocoon, feeling all round with its point, groping about; then, suddenly, the boring needle is released from its sheath, which falls back along the body, while the needle strives to make its entrance. The operation is a difficult one. I see the insect make a score of attempts, one after the other, without succeeding in piercing the tough wrapper of the Stelis. Should the instrument not penetrate, it retreats into its sheath and the insect resumes its scrutiny of the cocoon, sounding it point by point with the tips of its antennae. Then further thrusts are tried until one succeeds.

The eggs are little spindles, white and gleaming like ivory, about two-thirds of a millimeter in length. They have not the long, curved peduncle of the Leucospis' eggs; they are not suspended from the ceiling of the cocoon like these, but are laid without order around the fostering larva. Lastly, in a single cell and with a single mother, there is always more than one laying; and the number of eggs varies considerably in each. The Leucospis, because of her great size, which rivals that of her victim, the Bee, finds in each cell provisions enough for one and one alone. When, therefore, there is more than one set of eggs in any one cell, this is due to a mistake on her part and not a premeditated result. Where the whole ration is required for the meals of a single grub, she would take good care not to install several if she could help it. Her competitor is not called upon to observe the same discretion. A Chalicodoma grub gives the dwarf the wherewithal to portion a score of her little ones, who will live in common and in all comfort on what a single son of the giantess would eat up by himself. The tiny boring engineer, therefore, always settles a numerous family at the same banquet. The bowl, ample for a dozen or two, is emptied in perfect harmony.

Curiosity made me count the brood, to see if the mother was able to estimate the victuals and to proportion the number of guests to the sumptuousness of the fare provided. My notes mention fifty-four larvae in the cell of a masked Anthophora (Anthophora personata). No other census attained this figure. Possibly, two different mothers had laid their eggs in this crowded habitation. With the Mason bee of the Walls, I see the number of larvae vary, in different cells, between four and twenty-six; with the mason bee of the Sheds, between five and thirty-six; with the three-horned Osmia, who supplied me with the largest number of records, between seven and twenty-five; with the blue Osmia (Osmia cyanea, KIRB.), between five and six; with the Stelis (Stelis nasuta), between four and twelve.

The first return and the last two seem to point to some relation between the abundance of provisions and the number of consumers. When the mother comes upon the bountiful larva of the masked Anthophora, she gives it half-a-hundred to feed; with the Stelis and the blue Osmia, niggardly rations both, she contents herself with half-a-dozen. To introduce into the dining room only the number of boarders that the bill of fare will allow would certainly be a most deserving performance, especially as the insect is placed under very difficult conditions to judge the contents of the cell. These contents, which lie hidden under the ceiling, are invisible; and the insect can derive its information only from the outside of the nest, which varies in the different species. We should therefore have to admit the existence of a particular power of discrimination, a sort of discernment of the species, which is recognized as large or small from the outward aspect of its house. I refuse to go to this length in my conjectures, not that instinct seems to me incapable of such feats, but because of the particulars obtained from the three-horned Osmia and the two mason bees.

In the cells of these three species, I see the number of larvae put out to nurse vary in so elastic a fashion that I must abandon all idea of proportionate adjustment. The mother, without troubling unduly whether there be an excess or a dearth of provisions for her family, has filled the cells as her fancy prompted, or rather according to the number of ripe ovules contained in her ovaries at the time of the laying. If food be over-plentiful, the brood will be all the better for it and will grow bigger and stronger; if food be scarce, the famished youngsters will not die, but will remain smaller. Indeed, with both the larva and the full grown insect, I have often observed a difference in size which varies according to the density of the population, the members of a small colony being double the size of their overcrowded neighbors.

The grubs are white, tapering at both ends, sharply segmented and covered all over their bodies with a coat of fine, soft hairs which is invisible except under the lens. The head consists of a little knob much smaller in diameter than the body. In this head, the microscope reveals mandibles consisting of fine spikes of a tawny red, which spread into a wide, colorless base. Deprived of any indentation, incapable of chewing anything between their awl-shaped ends, these two tools serve at best to fix the grub slightly at some point of the fostering larva. Useless for carving, therefore, the mouth is a pure osculatory sucker, which drains the provisions by a process of exudation through the skin. We see here repeated what the Anthrax and the Leucospis have already shown us: the gradual exhaustion of a victim which the parasite consumes without killing it.

It is a curious spectacle even after that of the Anthrax. We have here twenty or thirty starvelings, all with their mouths pressed, as for a kiss, to the body of the plump larva, which, from day to day, fades and shrinks without the least appreciable wound, thus keeping fresh until reduced to a shriveled slough. If I disturb the gluttonous swarm, all, with a sudden recoil, let go, drop off and flounder around the foster mother. They are no less prompt in resuming their savage kisses. I need not add that neither at the point where they leave off nor at the point where they recommence is there the faintest trace of liquid. The oily exudation occurs only when the pump is at work. To linger over this strange method of feeding is superfluous after what I have said about the Anthrax.

The appearance of the full grown insect takes place at the beginning of summer, after nearly a whole year's stay in the invaded dwelling. The large number of inhabitants of one and the same cell led me to think that the work of deliverance ought to present a certain interest. They are all equally anxious to clear the walls of the prison at the earliest possible moment and to come forth into the great festival of the sun: do they all at the same time, in a confused horde, attack the ceiling which has to be pierced? Is the work of deliverance arranged in the general interest? Or is individual selfishness the only rule? These are the questions which observation will answer.

A little in advance of the proper season, I transfer each family into a short glass tube, which will represent the natal cell. A good, thick cork, quite a centimeter deep, is the obstacle to be pierced for an outlet. Well, instead of the mad haste and the ruinous lack of organization which I expected to find, my broods show me in their glass prison an exceedingly well regulated workshop. One insect, one only, works at perforating the cork. Patiently, with its mandibles, grain by grain, it digs a tunnel the width of its body. The gallery is so narrow that, in order to return to the tube, the worker has to move backwards. It is a slow process; and it takes hours and hours to dig the hole, a hard job for the frail miner.

Should her fatigue become too great, the excavator leaves the forefront and mingles with the crowd, to polish and dust herself. Another, the first neighbor at hand, at once takes her place and is herself relieved by a third when her task is done. Others again take their turn, always one at a time, so much so that the works are never at a standstill and never overcrowded. Meanwhile, the multitude keeps out of the way, quietly and patiently. There is no anxiety as to the deliverance. Success will come: of that they are all convinced. While waiting, one washes her antennae by passing them through her mouth, another polishes her wings with her hind legs, another frisks about to while away the period of inaction. Some are making love, a sovran means of killing time, whether one be born that day or twenty years ago.

Some, I said, make love. These favored ones are rare; they hardly count. Is it through indifference? No, but the gallants are lacking. The sexes are very unequally represented in the population of a cell: the males are in a wretched minority and sometimes even completely absent. This poverty did not escape the older observers. Brulle [Gaspard August Brulle (1809-1873)], the author of many works on natural history and one of the founders of the Societe entomologique de France, the only author whom I am able to consult in my hermitage, says, literally: 'The males do not appear to be known.'

I, for my part, know them; but, considering their feeble number, I keep asking myself what part they play in a harem so disproportionate to their forces. A few figures will show us what my hesitations are based upon.

In twenty-two Osmia cocoons (Osmia tricornis), the total census of the inmates yields three hundred and fifty-four, of whom forty-seven are males and three hundred and seven females. The average number of inmates, therefore, is sixteen individuals; and there are six females at least to one male. This disparity is maintained, in more or less marked proportions, whatever the species of the bee invaded. In the cocoons of the Mason bee of the Sheds, I discover the average proportion to be six females to one male; in those of the Mason bee of the Walls, I find one male to fifteen females.

These facts, which I am unable to state with any greater precision, are enough to give rise to the suspicion that the males, who are even tinier dwarfs than the females and who, moreover, like all insects, are injured by a single act of pairing, must, in most cases, remain strangers to the females. Can the mothers, in fact, dispense with their assistance, without being deprived of offspring on that account? I do not say yes, but I do not say no. The duality of the sexes is a hard problem. Why two sexes? Why not just one? It would have been much simpler and saved a great deal of foolery. Why such a thing as sex, when the tuber of the Jerusalem artichoke can do without it? These are the pregnant questions suggested to me, in the end, by Monodontomerus cupreus, the insect so infinitesimal in body and so overpowering in name that I had really vowed never to speak of it again by its official designation.

CHAPTER IV. LARVAL DIMORPHISM

If the reader has paid any attention to the story of the Anthrax, he must have perceived that my narrative is incomplete. The fox in the fable saw how the lion's visitors entered his den, but did not see how they went out. With us, it is the converse: we know the way out of the mason bee's fortress, but we do not know the way in. To leave the cell of which he has eaten the owner, the Anthrax becomes a perforating machine, a living tool from which our own industry might take a hint if it required new drills for boring rocks. When the exit tunnel is opened, this tool splits like a pod bursting in the sun; and from the stout framework there escapes a dainty fly, a velvety flake, a soft fluff that astounds us by its contrast with the roughness of the depths whence it ascends. On this point, we know pretty well what there is to know. There remains the entrance into the cell, a puzzle that has kept me on the alert for a quarter of a century.

To begin with, it is evident that the mother cannot lodge her egg in the cell of the mason bee, which has been long closed and barricaded with a cement wall by the time that the Anthrax makes her appearance. To penetrate it, she would have to become an excavating tool once more and resume the cast-off rags which she left behind in the exit window; she would have to retrace her steps, to be reborn a pupa; and life knows none of these retrogressions. The full grown insect, if endowed with claws, mandibles and plenty of perseverance, might at a pinch force the mortar casket; but the fly is not so endowed. Her slender legs would be strained and deformed by merely sweeping away a little dust; her mouth is a sucker for gathering the sugary exudations of the flowers and not the solid pincers needed for the crumbling of cement. There is no auger either, no bore copied from that of the Leucospis, no implement of any kind that can work its way into the thickness of the wall and dispatch the egg to its destination. In short, the mother is absolutely incapable of settling her eggs in the chamber of the Mason bee.

Can it be the grub that makes its own way into the storeroom, that same grub which we have seen draining the Chalicodoma with its leech-like kisses? Let us call the creature to mind: a little oily sausage, which stretches and curls up just where it lies, without being able to shift its position. Its body is a smooth cylinder; its mouth simply a circular lip. Not one ambulatory organ does it possess; not even hairs, protuberances or wrinkles to enable it to crawl. The animal is made for digestion and immobility. Its organization is incompatible with movement; everything tells us so in the clearest fashion. No, this grub is even less able than the mother to make its way unaided into the mason's dwelling. And yet the provisions are there; those provisions must be reached: it is a matter of life or death; to be or not to be. Then how does the fly set about it? It would be vain for me to question probabilities, too often illusory; to obtain a reply of any value, I have but one resource; I must attempt the nearly impossible and watch the Anthrax from the egg onwards.

Although Anthrax flies are fairly common, in the sense of there being several different species, they are not plentiful when it is a case of wanting a colony populous enough to admit of continuous observation. I see them, now here, now there, in the fiercely sun-scorched places, flitting hither and thither on the old walls, the slopes and the sand, sometimes in small platoons, most often singly. I can expect nothing of those vagabonds, who are here today and gone tomorrow, for I know nothing of their settlements. To keep a watch on them, one by one, in the blazing heat, is very painful and very unfruitful, as the swift-winged insect has a habit of disappearing one knows not whither just when a prospect of capturing its secret begins to offer. I have wasted many a patient hour at this pursuit, without the least result.

There might be some chance of success with Anthrax flies whose home was known to us beforehand, especially if insects of the same species formed a pretty numerous colony. The inquiries begun with one would be continued with a second and with more, until a complete verdict was forthcoming. Now, in the course of my long entomological career, I have met with but two species of Anthrax that fulfilled this condition and were to be found regularly: one at Carpentras; the other at Serignan. The first, Anthrax sinuata, FALLEN, lives in the cocoons of Osmia tricornis, who herself builds her nest in the old galleries of the hairy-footed Anthophora; the second, Anthrax trifasciata, MEIGEN, exploits the Chalicodoma of the Sheds. I will consult both.

Once more, here am I, somewhat late in life, at Carpentras, whose rude Gallic name sets the fool smiling and the scholar thinking. Dear little town where I spent my twentieth year and left the first bits of my fleece upon life's bushes, my visit of today is a pilgrimage; I have come to lay my eyes once more upon the place which saw the birth of the liveliest impressions of my early days. I bow, in passing, to the old college where I tried my prentice hand as a teacher. Its appearance is unchanged; it still looks like a penitentiary. Those were the views of our mediaeval educational system. To the gaiety and activity of boyhood, which were considered unwholesome, it applied the remedy of narrowness, melancholy and gloom. Its houses of instruction were, above all, houses of correction. The freshness of Virgil was interpreted in the stifling atmosphere of a prison. I catch a glimpse of a yard between four high walls, a sort of bear pit, where the scholars fought for room for their games under the spreading branches of a plane tree. All around were cells that looked like horse boxes, without light or air; those were the classrooms. I speak in the past tense, for doubtless the present day has seen the last of this academic destitution.

Here is the tobacco shop where, on Wednesday evening, coming out of the college, I would buy on credit the wherewithal to fill my pipe and thus to celebrate on the eve the joys of the morrow, that blessed Thursday [the weekly half-holiday in French schools] which I considered so well employed in solving hard equations, experimenting with new chemical reagents, collecting and identifying my plants. I would make my timid request, pretending to have come out without my money, for it is hard for a self-respecting man to admit that he is penniless. My candor appears to have inspired some little confidence; and I obtained credit, an unprecedented thing, with the representative of the revenue. [The government in France has the sole control of the tobacco trade, which forms an important branch of the inland revenue.] Ah, why did not I open a shop and expose for sale some packets of candles, a dozen dried cod, a barrel of sardines and a few cakes of soap! I am no more of a fool nor any less industrious than another; and I should have made my way. But, as it was, what could I expect? As an accoucheur of brains, a molder of intellects, I had no claim even to bread and cheese.

Here is my former habitation, occupied since by droning monks. In the embrasure of that window, sheltered from profane hands, between the closed outer shutters and the panes, I used to keep my chemicals, bought for a few sous cheated out of the weekly budget in the early days of our housekeeping. The bowl of a pipe was my crucible, a sweet jar my retort, mustard pots my receptacles for oxides and sulfides. My experiments, harmless or dangerous, were made on a corner of the fire beside the simmering broth.

How I should love to see that room again where I pored over differentials and integrals, where I calmed my poor burning head by gazing at Mont Ventoux, whose summit held in store for my coming expedition' those denizens of arctic climes, the saxifrage and the poppy! And to see my familiar friend, the blackboard which I hired at five francs a year from a crusty joiner, that board whose value I paid many times over, though I. could never buy it outright, for want of the necessary cash! The conic sections which I described on that blackboard, the learned hieroglyphics!

Though all my efforts, which were the more deserving because I had to work alone, led to almost nothing in that congenial calling, I would begin it all over again if I could. I should love to be conversing for the first time with Leibnitz and Newton, with Laplace and Lagrange, with Cuvier and

Jussieu, even if I had afterwards to solve that other arduous problem: how to procure one's daily bread. Ah, young men, my successors, what an easy time you have of it today! If you don't know it, then let me tell you so by means of these few pages from the life of one of your elders.

But let us not forget our insects, while listening to the echoes of illusions and difficulties roused in my memories by the cupboard window and the hired blackboard. Let us go back to the sunken roads of the Legue, which have become classic, so they say, since the appearance of my notes on the Oil beetles. Ye illustrious ravines, with your sun-baked slopes, if I have contributed a little to your fame, you, in your turn, have given me many fair hours of forgetfulness in the happiness of learning. You, at least, did not lure me with vain hopes; all that you promised you gave me and often a hundredfold. You are my promised land, where I would have sought at the last to pitch my observer's tent. My wish was not to be realized. Let me, at least, in passing, greet my beloved animals of the old days.

I raise my hat to Cerceris tuberculata, whom I see engaged on that slant, storing her Cleonus [a large species of weevil]. As I saw her then, so I see her now: the same staggering attempts to hoist the prey to the mouth of the burrow; the same brawls between males watching in the brushwood of the kermes oak. The sight of them sends a younger blood coursing through my veins; I receive as it were the breath of a new springtime of life. Time presses; let us pass on.

Another bow on this side. I hear buzzing up above, on that ledge, a colony of Sphex wasps, stabbing their crickets. We will give them a friendly glance, but no more. My acquaintances here are too numerous; I have not the leisure to renew my former relations with all of them. Without stopping, a wave of the hat to the Philanthi [bee-hunting wasps] who send the long avalanches of rubbish streaming down from their nests; and to Stizus ruficornis, [a hunting wasp] who stacks her praying mantises between two flakes of sandstone; and to the silky Ammophila [a digger wasp] with the red legs, who collects an underground store of loopers [also known as measuring worms, the larvae or caterpillars of the geometrid moth] and to the Tachtyti [hunting wasps], devourers of locusts; and to the Eumenes, builders of clay cupolas on a bough.

Here we are at last. This high, perpendicular rock, facing the south to a length of some hundreds of yards and riddled with holes like a monstrous sponge, is the time-honored dwelling place of the hairy-footed Anthophora and of her rent free tenant, the three-horned Osmia. Here also swarm their exterminators: the Sitaris beetle, the parasite of the Anthophora; the Anthrax fly, the murderer of the Osmia. Ill informed as to the proper period, I have come rather late, on the 10th of September. I should have been here a month ago, or even by the end of July, to watch the fly's operations. My journey threatens to be fruitless: I see but a few rare Anthrax flies, hovering round the face of the cliff. We will not despair, however, and we will begin by consulting the locality.

The Anthophora's cells contain this bee in the larval stage. Some of them provide me with the oil beetle and the Sitaris, rare finds at one time, today of no use to me. Others contain the Melecta [a parasitic bee] in the form of a highly colored pupa, or even in that of the full grown insect. The Osmia, still more precocious, though dating from the same period, shows herself exclusively in the adult form, a bad omen for my investigations, for what the Anthrax demands is the larva and not the perfect insect. The fly's grub doubles my apprehensions. Its development is complete, the larva on which it feeds is consumed, perhaps several weeks ago. I no longer doubt but that I have come too late to see what happens in the Osmia's cocoons.

Is the game lost? Not yet. My notes contain evidence of Anthrax flies hatching in the latter half of September. Besides, those whom I now see exploring the rock are not there to take exercise: their preoccupation is the settling of the family. These belated ones cannot tackle the Osmia, who, with her firm, adult flesh, would not suit the nursling's delicate needs and who, moreover, powerful as she is, would offer resistance. But in autumn a less numerous colony of honey gatherers takes the place, upon the slope, of the spring colony, from which it differs in species. In particular, I see the Diadem Anthidium [a clothier bee who lines her nest with wool and cotton] at work, entering her galleries at one time with her harvest of pollen dust and at another with her little bale of cotton. Might not these autumnal Bees be themselves exploited by the Anthrax, the same that selected the Osmia as her victim a couple of months earlier? This would explain the presence of the Anthrax flies whom I now see fussing about.

A little reassured by this conjecture, I take my stand at the foot of the rock, under a broiling sun; and, for half a day, I follow the evolutions of my flies. They flit quietly in front of the slope, at a few inches from the earthy covering. They go from one orifice to the next, but without even penetrating. For that matter, their big wings, extended crosswise even when at rest, would resist their entrance into a gallery, which is too narrow to admit those spreading sails. And so they explore the cliff, going to and fro and up and down, with a flight that is now sudden, now smooth and slow. From time to time, I see the Anthrax quickly approach the wall and lower her abdomen as though to touch the earth with the end of her ovipositor. This proceeding takes no longer than the twinkling of an eye. When it is done, the insect alights elsewhere and rests. Then it resumes its sober flight, its long investigations and its sudden blows with the tip of its belly against the layer of earth. The Bombylii [bee flies] observe similar tactics when soaring at a short height above the ground.

I at once rushed to the spot touched, lens in hand, in the hope of finding the egg which everything told me was laid during that tap of the abdomen. I could distinguish nothing, in spite of the closest attention. It is true that my exhaustion, together with the blinding light and scorching heat, made examination very difficult. Afterwards, when I made the acquaintance of the tiny thing that issues from that egg, my failure no longer surprised me. In the leisure of my study, with my eyes rested and with my most powerful glasses held in a hand no longer shaking with excitement and fatigue, I have the very greatest difficulty in finding the infinitesimal creature, though I know exactly where it lies. Then how could I see the egg, worn out as I was under the sun-baked cliff, how discover the precise spot of a laying performed in a moment by an insect seen only at a distance? In the painful conditions wherein I found myself, failure was inevitable.

Despite my negative attempts, therefore, I remain convinced that the Anthrax flies strew their eggs one by one, on the spots frequented by those bees who suit their grubs. Each of their sudden strokes with the tip of the abdomen represents a laying. They take no precaution to place the germ under cover; for that matter, any such precaution would be rendered impossible by the mother's structure. The egg, that delicate object, is laid roughly in the blazing sun, between grains of sand, in some wrinkle of the calcined chalk. That summary installation is sufficient, provided the coveted larva be near at hand. It is for the young grub now to manage as best it can at its own risk and peril.

Though the sunken roads of the Legue did not tell me all that I wished to know, they at least made it very probable that the coming grub must reach the victualled cell by its own efforts. But the grub which we know, the one that drains the bag of fat which may be a Chalicodoma larva or an Osmia larva, cannot move from its place, still less indulge in journeys of discovery through the thickness of a wall and the web of a cocoon. So an imperative necessity presents itself: there must perforce be an initial larva form, capable of moving and organized for searching, a form under which the grub would attain its end. The Anthrax would thus possess two larval states: one to penetrate to the provisions; the other to consume them. I allow myself to be convinced by the logic of it all; I already see in my mind's eye the wee animal coming out of the egg, endowed with sufficient power of motion not to dread a walk and with sufficient slenderness to glide into the smallest crevices. Once in the presence of the larva on which it is to feed, it doffs its travelling dress and becomes the obese animal whose one duty it is to grow big and fat in immobility. This is all very coherent; it is all deduced like a geometrical proposition. But to the wings of imagination, however smooth their flight, we must prefer the sandals of observed facts, the slow sandals with the leaden soles. Thus shod, I proceed.

Next year, I resume my investigations, this time on the Anthrax of the Chalicodoma, who is my neighbor in the surrounding wastelands and will allow me to repeat my visits daily, morning and evening if need be. Taught by my earlier studies, I now know the exact period of the Bee's hatching and therefore of the Anthrax' laying, which must take place soon after. Anthrax trifasciata settles her family in July, or in August at latest. Every morning, at nine o'clock, when the heat begins to be unendurable and when, to use [the author's gardener and factotum] Favier's expression, an extra log is flung on the bonfire of the sun, I take the field, prepared to come back with my head aching from the glare, provided that I bring home the solution of my puzzle. A man must have the devil in him to leave the shade at this time of the year. And what for, pray? To write the story of a fly! The greater the heat, the better my chance of success. What causes me to suffer torture fills the insect with delight; what prostrates me braces the fly. Come along!

The road shimmers like a sheet of molten steel. From the dusty and melancholy olive trees rises a mighty, throbbing hum, a great andante whose executants have the whole sweep of woods for their orchestra. 'Tis the concert of the Cicada, whose bellies sway and rustle with increasing frenzy as the temperature rises. The strident scrapings of the Cicada of the Ash, the Carcan of the district, lend their rhythm to the one note symphony of the common cicada. This is the moment: come along! And, for five or six weeks, oftenest in the morning, sometimes in the afternoon, I set myself to explore the flinty plateau.

The Chalicodoma's nests abound, but I cannot see a single Anthrax make a black speck upon their surface. Not one, busy with her laying, settles in front of me. At most, from time to time, I can just see one passing far away, with an impetuous rush. I lose her in the distance; and that is all. It is impossible to be present at the laying of the egg. I know the little that I learnt from the cliffs in the Legue and nothing more.

As soon as I recognize the difficulty, I hasten to enlist assistants. Shepherds—mere small boys—keep the sheep in these stony meadows, where the flocks graze, to the greater glory of our local mutton, on the camphor saturated badafo, that is to say, spike lavender. I explain as well as I can the object of my search; I talk to them of a big black Fly and the nests on which she ought to settle, the clay nests so well known to those who have learnt how to extract the honey with a straw in springtime and spread it on a crust of bread. They are to watch that fly and take good note of the nests on which they may see her alight; and, on the same evening, when they bring their flocks back to the village, they are to tell me the result of their day's work. On receiving their favorable report, I will go with them, next day, to continue the observations. They shall be paid for their trouble, of course. These latter day Corydons have not the manners of antiquity: they reck little of the seven holed flute cemented with wax, or of the beechen bowl, preferring the coppers that will take them to the village inn on Sunday. A reward in ready money is promised for each nest that fulfils the desired conditions; and the bargain is enthusiastically accepted.

There are three of them; and I make a fourth. Shall we manage it, among us all? I thought so. By the end of August, however, my last illusions were dispelled. Not one of us had succeeded in seeing the big black Fly perching on the dome of the mason bee.

Our failure, it seems to me, can be explained thus: outside the spacious front of the Anthophora's settlement, the Anthrax is in permanent residence. She visits, on the wing, every nook and corner, without moving away from the native cliff, because it would be useless to go farther. There is board and lodging here, indefinitely, for all her family. When some spot is deemed favorable, she hovers round inspecting it, then comes up suddenly and strikes it with the tip of her abdomen. The thing is done, the egg is laid. So I picture it, at least. Within a radius of a few yards and in a flight broken by short intervals of rest in the sun, she carries on her search of likely places for the laying and dissemination of her eggs. The insect's assiduous attendance upon the same slope is caused by the inexhaustible wealth of the locality exploited.

The Anthrax of the Chalicodoma labors under very different conditions. Stay-at-home habits would be detrimental to her. With her rushing flight, made easy by the long and powerful spread of her wings, she must travel far and wide if she would found a colony. The bee's nests are not discovered in groups, but occur singly on their pebbles, scattered more or less everywhere over acres of ground. To find a single one is not enough for the fly: on account of the many parasites, not all the cells, by a long way, contain the desired larva; others, too well protected, would not allow of access to the provisions. Very many nests are necessary, perhaps, for the eggs of one alone; and the finding of them calls for long journeys.

I therefore picture the Anthrax coming and going in every direction across the stony plain. Her practiced eye requires no slackened flight to distinguish the earthen dome which she is seeking. Having found it, she inspects it from above, still on the wing; she taps it once and yet once again with the tip of her ovipositor and forthwith makes off, without having set foot on the ground. Should she take a rest, it will be elsewhere, no matter where, on the soil, on a stone, on a tuft of lavender or thyme. Given these habits—and my observations in the Carpentras roads make them seem exceedingly probable—it is small wonder that the perspicacity of my young shepherds and myself should have come to naught. I was expecting the impossible: the Anthrax does not halt on the mason bee's nest to proceed with her laying in a methodical fashion; she merely pays a flying visit.

And so I develop my theory of a primary larval form, differing in every way from the one which I know. The organization of the Anthrax must be such, at the beginning, as to permit of its moving on the surface of the dome where the egg has been dropped so carelessly; the nascent grub must be supplied with tools to pierce the concrete wall and enter the Bee's cell through some cranny. The fly grub, perhaps dragging the remnants of the egg behind it, must set out in quest of board and lodging almost as soon as it is born. It will succeed under the guidance of instinct, that faculty which waits not to number the days and which is as far seeing at the moment of hatching as after the trials of a busy life. This primary grub does not seem to me outside the limits of possibility; I see it, if not in the body, at least in its actions, as plainly as though it were really under the lens. It exists, if reason be not a vain and empty guide; I must find it; I shall find it. Never in the history of my investigations has the logic of things been more insistent; never has it directed me with greater certainty towards a magnificent biological theory.

While vainly trying to witness the laying of the eggs, I inquire, at the same time, into the contents of the Mason bee's nests, in quest of the grub just issued from the egg. My own harvest and that of my young shepherds, whose zeal I employ in a task less difficult than the first, procure me heaps of nests, enough to fill baskets and baskets. These are all inspected at leisure, on my work table, with the excitement which the certainty of an approaching fine discovery never fails to give. The Mason's cocoons are taken from the cells, inspected without, opened and inspected within. My lens explores their innermost recesses; speck by speck, it explores the Chalicodoma's slumbering larva; it explores the inner walls of the cells. Nothing, nothing, nothing! For a fortnight and more, nests were rejected and heaped up in a corner; my study was crammed with them. What hecatombs of unfortunate sleepers removed from their silken bags and doomed, for the most part, to a wretched end, despite the care which I took to put them in a place of safety, where the work of the transformation might be pursued! Curiosity makes us cruel. I continue to rip up cocoons. And nothing, nothing! It needed the sturdiest faith to make me persevere. That faith I possessed; and well for me that I did.

On the 25th of July—the date deserves to be recorded—I saw, or rather seemed to see, something move on the Chalicodoma's larva. Was it an illusion born of my hopes? Was it a bit of diaphanous down stirred by my breath? It was not an illusion, it was not a bit of down, it was really and truly a grub. What a moment, followed by what perplexities! The thing has nothing in common with the larva of the Anthrax, it suggests rather some microscopic Thread worm that, by accident, has made its way through the skin of its host and come to enjoy itself outside. I do not reckon my discovery as of much value, because I am so greatly puzzled by the creature's appearance. No matter: we will take a small glass tube and place inside it the Chalicodoma grub and the mysterious thing wriggling on the surface. Suppose it should be what I am looking for? Who knows?

Once warned of the probable difficulty of seeing the animalcule for which I am hunting, I redouble my attention, so much so that, in a couple of days, I am the owner of half a score of tiny worms similar to the one which caused me such excitement. Each of them is lodged in a glass tube with its Chalicodoma grub. The infinitesimal thing is so small, so diaphanous, blends to such good purpose with its host that the least fold of skin conceals it from my view. After watching it one day through the lens, I sometimes fail to find it again on the morrow. I think that I have lost it, that it has perished under the weight of the overturned larva and returned to that nothing to which it was so closely akin. Then it moves and I see it again. For a whole fortnight, there was no limit to my perplexity. Was it really the original larva of the Anthrax? Yes, for I at last saw my bantlings transform themselves into the larva previously described and make their first start at draining their victims with kisses. A few moments of satisfaction like those which I then enjoyed make up for many a weary hour.

Let us resume the story of the wee animal, now recognized as the genuine origin of the Anthrax. It is a tiny worm about a millimeter long and almost as slender as a hair. It is very difficult to see because of its transparency. When tucked away in a fold of the skin of its fostering larva, an excessively fine skin, it remains undiscoverable to the lens. But the feeble creature is very active: it tramps over the sides of the rich morsel, walks all round it. It covers the ground pretty quickly, buckling and unbuckling by turns, very much after the manner of the looper caterpillar. Its two extremities are its chief points of support. When at a standstill, it moves its front half in every direction, as though to explore the space around it; when walking, it swells out, magnifies its segments and then looks like a bit of knotted string.

The microscope shows us thirteen rings, including the head. This head is small, slightly horny, as is proved by its amber color, and bristles in front with a small number of short, stiff hairs. On each of the three segments of the thorax there are two long hairs, fixed to the lower surface; and there are two similar and still longer hairs at the end of the terminal ring. These four pairs of bristles, three in front and one behind, are the locomotory organs, to which we must add the hairy edge of the head and also the anal button, a sustaining base which might very well work with the aid of a certain stickiness, as happens with the primary larva of the Sitaris [a Parasitic Beetle noted for the multiplicity of transformations undergone by the grub]. We see, through the transparent skin, two long air tubes running parallel to each other from the first thoracic segment to the last abdominal segment but one. They ought to end in two pairs of breathing holes which I have not succeeded in distinguishing quite plainly. Those two big respiratory vessels are characteristic of the grubs of flies. Their mouths correspond exactly with the points at which the two sets of stigmata open in the Anthrax larva in its second form.

For a fortnight, the feeble grub remains in the condition which I have described, without growing and very probably also without nourishment. Assiduous though my visits be, I never perceive it taking any refreshment. Besides, what would it eat? In the cocoon invaded there is nothing but the larva of the mason bee; and the worm cannot make use of this before acquiring the sucker that comes with the second form. Nevertheless, this life of abstinence is not a life of idleness. The animalcule explores its dish, now here, now elsewhere; it runs all over it with looper strides; it pries into the neighborhood by lifting and shaking its head.

I see a need for this long wait under a transitory form that requires no feeding. The egg is laid by the mother on the surface of the nest, somewhere near a suitable cell, I dare say, but still at a distance from the fostering larva, which is protected by a thick rampart. It is for the new born grub to make its own way to the provisions, not by violence and house breaking, of which it is incapable, but by patiently slipping through a maze of cracks, first tried, then abandoned, then tried again. It is a very difficult task, even for this most slender worm, for the bee's masonry is exceedingly compact. There are no chinks due to bad building; no fissures due to the weather; nothing but an apparently impenetrable homogeneity. I see but one weak part and that only in a few nests: it is the line where the dome joins the surface of the stone. An imperfect soldering between two materials of different nature, cement and flint, may leave a breach wide enough to admit besiegers as thin as a hair. Nevertheless, the lens is far from always finding an inlet of this kind on the nests occupied by Anthrax flies.

And so I am ready to allow that the animalcule wandering in search of its cell has the whole area of the dome at its disposal when selecting an entrance. Where the line auger of the Leucospis can enter, is there not room enough for the even slimmer Anthrax grub? True, the Leucospis possesses muscular force and a hard boring tool. The Anthrax is extremely weak and has nothing but invincible patience. It does at great length of time what the other, furnished with superior implements, accomplishes in three hours. This explains the fortnight spent by the Anthrax under the initial form, the object of which is to overcome the obstacle of the mason's wall, to pierce through the texture of the cocoon and to reach the victuals.

I even believe that it takes longer. The work is so laborious and the worker so feeble! I cannot tell how long it is since my bantlings attained their object. Perhaps, aided by easy roads, they had reached their fostering larvae long before the completion of their first babyhood, the end of which they were spending before my eyes, with no apparent purpose, in exploring their provisions. The time had not yet come for them to change their skins and take their seats at the table. Their fellows must still, for the most part, be wandering through the pores of the masonry; and this was what made my search so vain at the start.

A few facts seem to suggest that the entrance into the cell may be delayed for several months by the difficulty of the passages. There are a few Anthrax grubs beside the remains of pupae not far removed from the final metamorphosis; there are others, but very rarely, on Mason bees already in the perfect state. These grubs are sickly and appear to be ailing; the provisions are too solid and do not lend themselves to the delicate suckling of the worms. Who can these laggards be but animalcules that have roamed too long in the walls of the nest? Failing to make their entrance at the proper time, they no longer find viands to suit them. The primary larva of the Sitaris continues from the autumn to the following spring. Even so the initial form of the Anthrax might well continue, not in inactivity, but in stubborn attempts to overcome the thick bulwark.

My young worms, when transferred with their provisions into tubes, remained stationary, on the average, for a couple of weeks. At last, I saw them shrink and then rid themselves of their epidermis and become the grub which I was so anxiously expecting as the final reply to all my doubts. It was indeed, from the first, the grub of the Anthrax, the cream-colored cylinder with the little button of a head, followed by a hump. Applying its cupping glass to the mason bee, the worm, without delay, began its meal, which lasts another fortnight. The reader knows the rest.

Before taking leave of the animalcule, let us devote a few lines to its instinct. It has just awakened to life under the fierce kisses of the sun. The bare stone is its cradle, the rough clay its welcomer, as it makes its entrance into the world, a poor thread of scarce cohering albumen. But safety lies within; and behold the atom of animated glair embarking on its struggle with the flint. Obstinately, it sounds each pore; it slips in, crawls on, retreats, begins again. The radical of the germinating seed is no more persevering in its efforts to descend into the cool earth than is the Anthrax grub in creeping into the lump of mortar. What inspiration urges it towards its food at the bottom of the clod, what compass guides it? What does it know of those depths, of what lies therein or where? Nothing. What does the root know of the earth's fruitfulness? Again nothing. Yet both make for the nourishing spot. Theories are put forward, most learned theories, introducing capillary action, osmosis and cellular imbibition, to explain why the caulicle ascends and the radical descends. Shall physical or chemical forces explain why the animalcule digs into the hard clay? I bow profoundly, without understanding or even trying to understand. The question is far above, our inane means.

The biography of the Anthrax is now complete, save for the details relating to the egg, as yet unknown. In the vast majority of insects subject to metamorphoses, the hatching yields the larval form which will remain unchanged until the nymphosis. By virtue of a remarkable variation, revealing a new vein of observation to the entomologist, the Anthrax flies, in the larval state, assume two successive shapes, differing greatly one from the other, both in structure and in the part which they are called upon to play. I will describe this double stage of the organism by the phrase 'larval dimorphism.' The initial form, that issuing from the egg, I will call 'the primary larva;' the second form shall be 'the secondary larva.' Among the Anthrax flies, the function of the primary larva is to reach the provisions, on which the mother is unable to lay her egg. It is capable of moving and endowed with ambulatory bristles, which allow the slim creature to glide through the smallest interstices in the wall of a Bee's nest, to slip through the woof of the cocoon and to make its way to the larva intended for its successor's food. When this object is attained, its part is played. Then appears the secondary larva, deprived of any means of progression. Relegated to the inside of the invaded cell, as incapable of leaving it by its own efforts as it was of entering, this one has no mission in life but that of eating. It is a stomach that loads itself, digests and goes on adding to its reserves. Next comes the pupa, armed for the exit even as the primary larva was equipped for entering. When the deliverance is accomplished, the perfect insect appears, busy with its laying. The Anthrax cycle is thus divided into four periods, each of which corresponds with special forms and functions. The primary larva enters the casket containing provisions; the secondary larva consumes these provisions; the pupa brings the insect to light by boring through the enclosing wall; the perfect insect strews its eggs; and the cycle starts afresh.

CHAPTER V. HEREDITY

Facts which I have set forth elsewhere prove that certain dung beetles' make an exception to the rule of paternal indifference—a general rule in the insect world—and know something of domestic cooperation. The father works with almost the same zeal as the mother in providing for the settlement of the family. Whence do these favored ones derive a gift that borders on morality?

One might suggest the cost of installing the youngsters. Once they have to be furnished with a lodging and to be left the wherewithal to live, is it not an advantage, in the interests of the race, that the father should come to the mother's assistance? Work divided between the two will ensure the comfort which solitary work, its strength overtaxed, would deny. This seems excellent reasoning; but it is much more often contradicted than confirmed by the facts. Why is the Sisyphus a hard working paterfamilias and the sacred beetle an idle vagabond? And yet the two pill rollers practice the same industry and the same method of rearing their young. Why does the Lunary Copris know what his near kinsman, the Spanish Copris, does not? The first assists his mate, never forsakes her. The second seeks a divorce at an early stage and leaves the nuptial roof before the children's rations are massed and kneaded into shape. Nevertheless, on both sides, there is the same big outlay on a cellarful of egg-shaped pills, whose neat rows call for long and watchful supervision. The similarity of the produce leads one to believe in similarity of manners; and this is a mistake.

Let us turn elsewhere, to the wasps and bees, who unquestionably come first in the laying up of a heritage for their offspring. Whether the treasure hoarded for the benefit of the sons be a pot of honey or a bag of game, the father never takes the smallest part in the work. He does not so much as give a sweep of the broom when it comes to tidying the outside of the dwelling. To do nothing is his invariable rule. The bringing up of the family, therefore, however expensive it may be in certain cases, has not given rise to the instinct of paternity. Then where are we to look for a reply?

Let us make the question a wider one. Let us leave the animal, for a moment, and occupy ourselves with man. We have our own instincts, some of which take the name of genius when they attain a degree of might that towers over the plain of mediocrity. We are amazed by the unusual, springing out of flat commonplaces; we are spellbound by the luminous speck shining in the wonted darkness. We admire; and, failing to understand whence came those glorious harvests in this one or in that, we say of them: "They have the gift."

A goatherd amuses himself by making combinations with heaps of little pebbles. He becomes an astoundingly quick and accurate reckoner without other aid than a moment's reflection. He terrifies us with the conflict of enormous numbers which blend in an orderly fashion in his mind, but whose mere statement overwhelms us by its inextricable confusion. This marvelous arithmetical juggler has an instinct, a genius, a gift for figures.

A second, at the age when most of us delight in tops and marbles, leaves the company of his boisterous playmates and listens to the echo of celestial harps singing within him. His head is a cathedral filled with the strains of an imaginary organ. Rich cadences, a secret concert heard by him and him alone, steep him in ecstasy. All hail to that predestined one who, some day, will rouse our noblest emotions with his musical chords. He has an instinct, a genius, a gift for sounds.

A third, a brat who cannot yet eat his bread and jam without smearing his face all over, takes a delight in fashioning clay into little figures that are astonishingly lifelike for all their artless awkwardness. He takes a knife and makes the briar root grin into all sorts of entertaining masks; he carves boxwood in the semblance of a horse or sheep; he engraves the effigy of his dog on sandstone. Leave him alone; and, if Heaven second his efforts, he may become a famous sculptor. He has an instinct, a gift, a genius for form.

And so with others in every branch of human activity: art and science, industry and commerce, literature and philosophy. We have within us, from the start, that which will distinguish us from the vulgar herd. Now to what do we owe this distinctive character? To some throwback of atavism, men tell us. Heredity, direct in one case, remote in another, hands it down to us, increased or modified by time. Search the records of the family and you will discover the source of the genius, a mere trickle at first, then a stream, then a mighty river.

The darkness that lies behind that word heredity! Metaphysical science has tried to throw a little light upon it and has succeeded only in making unto itself a barbarous jargon, leaving obscurity more obscure than before. As for us, who hunger after lucidity, let us relinquish abstruse theories to whoever delights in them and confine our ambition to observable facts, without pretending to explain the quackery of the plasma. Our method certainly will not reveal to us the origin of instinct; but it will at least show us where it would be waste of time to look for it.

In this sort of research, a subject known through and through, down to its most intimate peculiarities, is indispensable. Where shall we find that subject? There would be a host of them and magnificent ones, if it were possible to read the sealed pages of others' lives; but no one can sound an existence outside his own and even then he can think himself lucky if a retentive memory and the habit of reflection give his soundings the proper accuracy. As none of us is able to project himself into another's skin, we must needs, in considering this problem, remain inside our own.

To talk about one's self is hateful, I know. The reader must have the kindness to excuse me for the sake of the study in hand. I shall take the silent beetle's place in the witness box, cross-examining myself in all simplicity of soul, as I do the animal, and asking myself whence that one of my instincts which stands out above the others is derived.

Since Darwin bestowed upon me the title of 'incomparable observer,' the epithet has often come back to me, from this side and from that, without my yet understanding what particular merit I have shown. It seems to me so natural, so much within everybody's scope, so absorbing to interest one's self in everything that swarms around us! However, let us pass on and admit that the compliment is not unfounded.

My hesitation ceases if it is a question of admitting my curiosity in matters that concern the insect. Yes, I possess the gift, the instinct that impels me to frequent that singular world; yes, I know that I am capable of spending on those studies an amount of precious time which would be better employed in making provision, if possible, for the poverty of old age; yes, I confess that I am an enthusiastic observer of the animal. How was this characteristic propensity, at once the torment and delight of my life, developed? And, to begin with, how much does it owe to heredity?

The common people have no history: persecuted by the present, they cannot think of preserving the memory of the past. And yet what surpassingly instructive records, comforting too and pious, would be the family papers that should tell us who our forebears were and speak to us of their patient struggles with harsh fate, their stubborn efforts to build up, atom by atom, what we are today. No story would come up with that for individual interest. But by the very force of things the home is abandoned; and, when the brood has flown, the nest is no longer recognized.

I, a humble journeyman in the toilers' hive, am therefore very poor in family recollections. In the second degree of ancestry, my facts become suddenly obscured. I will linger over them a moment for two reasons: first, to inquire into the influence of heredity; and, secondly, to leave my children yet one more page concerning them.

I did not know my maternal grandfather. This venerable ancestor was, I have been told, a process server in one of the poorest parishes of the Rouergue. He used to engross on stamped paper in a primitive spelling. With his well-filled pen case and ink horn, he went drawing out deeds up hill and down dale, from one insolvent wretch to another more insolvent still. Amid his atmosphere of pettifoggery, this rudimentary scholar, waging battle on life's acerbities, certainly paid no attention to the insect; at most, if he met it, he would crush it under foot. The unknown animal, suspected of evil doing, deserved no further enquiry. Grandmother, on her side, apart from her housekeeping and her beads, knew still less about anything. She looked on the alphabet as a set of hieroglyphics only fit to spoil your sight for nothing, unless you were scribbling on paper bearing the government stamp. Who in the world, in her day, among the small folk, dreamt of knowing how to read and write? That luxury was reserved for the attorney, who himself made but a sparing use of it. The insect, I need hardly say, was the least of her cares. If sometimes, when rinsing her salad at the tap, she found a caterpillar on the lettuce leaves, with a start of fright she would fling the loathsome thing away, thus cutting short relations reputed dangerous. In short, to both my maternal grandparents, the insect was a creature of no interest whatever and almost always a repulsive object, which one dared not touch with the tip of one's finger. Beyond a doubt, my taste for animals was not derived from them.

I have more precise information regarding my grandparents on the father's side, for their green old age allowed me to know them both. They were people of the soil, whose quarrel with the alphabet was so great that they had never opened a book in their lives; and they kept a lean farm on the cold granite ridge of the Rouergue tableland. The house, standing alone among the heath and broom, with no neighbor for many a mile around and visited at intervals by the wolves, was to them the hub of the universe. But for a few surrounding villages, whither the calves were driven on fair days, the rest was only very vaguely known by hearsay. In this wild solitude, the mossy fens, with their quagmires oozing with iridescent pools, supplied the cows, the principal source of wealth, with rich, wet grass. In summer, on the short swards of the slopes, the sheep were penned day and night, protected from beasts of prey by a fence of hurdles propped up with pitchforks. When the grass was cropped close at one spot, the fold was shifted elsewhere. In the center was the shepherd's rolling hut, a straw cabin. Two watchdogs, equipped with spiked collars, were answerable for tranquillity if the thieving wolf appeared in the night from out the neighboring woods.

Padded with a perpetual layer of cow dung, in which I sank to my knees, broken up with shimmering puddles of dark brown liquid manure, the farmyard also boasted a numerous population. Here the lambs skipped, the geese trumpeted, the fowls scratched the ground and the sow grunted with her swarm of little pigs hanging to her dugs.

The harshness of the climate did not give husbandry the same chances. In a propitious season, they would set fire to a stretch of moorland bristling with gorse and send the swing plow across the ground enriched with the cinders of the blaze. This yielded a few acres of rye, oats and potatoes. The best corners were kept for hemp, which furnished the distaffs and spindles of the house with the material for linen and was looked upon as grandmother's private crop.

Grandfather, therefore, was, before all, a herdsman versed in matters of cows and sheep, but completely ignorant of aught else. How dumbfounded he would have been to learn that, in the remote future, one of his family would become enamoured of those insignificant animals to which he had never vouchsafed a glance in his life! Had he guessed that that lunatic was myself, the scapegrace seated at the table by his side, what a smack I should have caught in the neck, what a wrathful look!

"The idea of wasting one's time with that nonsense!" he would have thundered.

For the patriarch was not given to joking. I can still see his serious face, his unclipped head of hair, often brought back behind his ears with a flick of the thumb and spreading its ancient Gallic mane over his shoulders. I see his little three-cornered hat, his small clothes buckled at the knees, his wooden shoes, stuffed with straw, that echoed as he walked. Ah, no! Once childhood's games were past, it would never have done to rear the Grasshopper and unearth the Dung beetle from his natural surroundings.

Grandmother, pious soul, used to wear the eccentric headdress of the Rouergue highlanders: a large disk of black felt, stiff as a plank, adorned in the middle with a crown a finger's breadth high and hardly wider across than a six franc piece. A black ribbon fastened under the chin maintained the equilibrium of this elegant, but unsteady circle. Pickles, hemp, chickens, curds and whey, butter; washing the clothes, minding the children, seeing to the meals of the household: say that and you have summed up the strenuous woman's round of ideas. On her left side, the distaff, with its load of flax; in her right hand, the spindle turning under a quick twist of her thumb, moistened at intervals with her tongue: so she went through life, unwearied, attending to the order and the welfare of the house. I see her in my mind's eye particularly on winter evenings, which were more favorable to family talk. When the hour came for meals, all of us, big and little, would take our seats round a long table, on a couple of benches, deal planks supported by four rickety legs. Each found his wooden bowl and his tin spoon in front of him. At one end of the table always stood an enormous rye loaf, the size of a cartwheel, wrapped in a linen cloth with a pleasant smell of washing, and remained until nothing was left of it. With a vigorous stroke, grandfather would cut off enough for the needs of the moment; then he would divide the piece among us with the one knife which he alone was entitled to wield. It was now each one's business to break up his bit with his fingers and to fill his bowl as he pleased.

Next came grandmother's turn. A capacious pot bubbled lustily and sang upon the flames in the hearth, exhaling an appetizing savor of bacon and turnips. Armed with a long metal ladle, grandmother would take from it, for each of us in turn, first the broth, wherein to soak the bread, and next the ration of turnips and bacon, partly fat and partly lean, filling the bowl to the top. At the other end of the table was the pitcher, from which the thirsty were free to drink at will. What appetites we had and what festive meals those were, especially when a cream cheese, homemade, was there to complete the banquet!

Near us blazed the huge fireplace, in which whole tree trunks were consumed in the extreme cold weather. From a corner of that monumental, soot-glazed chimney, projected, at a convenient height, a bracket with a slate shelf, which served to light the kitchen when we sat up late. On this we burnt chips of pine wood, selected among the most translucent, those containing the most resin. They shed over the room a lurid red light, which saved the walnut oil in the lamp.

When the bowls were emptied and the last crumb of cheese scraped up, grandam went back to her distaff, on a stool by the chimney corner. We children, boys and girls, squatting on our heels and putting out our hands to the cheerful fire of furze, formed a circle round her and listened to her with eager ears. She told us stories, not greatly varied, it is true, but still wonderful, for the wolf often played a part in them. I should have very much liked to see this wolf, the hero of so many tales that made our flesh creep; but the shepherd always refused to take me into his straw hut, in the middle of the fold, at night. When we had done talking about the horrid wolf, the dragon and the serpent and when the resinous splinters had given out their last gleams, we went to sleep the sweet sleep that toil gives. As the youngest of the household, I had a right to the mattress, a sack stuffed with oat chaff. The others had to be content with straw.

I owe a great deal to you, dear grandmother: it was in your lap that I found consolation for my first sorrows. You have handed down to me, perhaps, a little of your physical vigor, a little of your love of work; but certainly you were no more accountable than grandfather for my passion for insects.

Nor was either of my own parents. My mother, who was quite illiterate, having known no teacher than the bitter experience of a harassed life, was the exact opposite of what my tastes required for their development. My peculiarity must seek its origin elsewhere: that I will swear. But I do not find it in my father, either. The excellent man, who was hard working and sturdily built like granddad, had been to school as a child. He knew how to write, though he took the greatest liberties with spelling; he knew how to read and understood what he read, provided the reading presented no more serious literary difficulties than occurred in the stories in the almanac. He was the first of his line to allow himself to be tempted by the town and he lived to regret it. Badly off, having but little outlet for his industry, making God knows what shifts to pick up a livelihood, he went through all the disappointments of the countryman turned townsman. Persecuted by bad luck, borne down by the burden, for all his energy and good will, he was far indeed from starting me in entomology. He had other cares, cares more direct and more serious. A good cuff or two when he saw me pinning an insect to a cork was all the encouragement that I received from him. Perhaps he was right.

The conclusion is positive: there is nothing in heredity to explain my taste for observation. You may say that I do not go far enough back. Well, what should I find beyond the grandparents where my facts come to a stop? I know, partly. I should find even more uncultured ancestors: sons of the soil, plowmen, sowers of rye, neat herds; one and all, by the very force of things, of not the least account in the nice matters of observation.

And yet, in me, the observer, the inquirer into things began to take shape almost in infancy. Why should I not describe my first discoveries? They are ingenuous in the extreme, but will serve notwithstanding to tell us something of the way in which tendencies first show themselves. I was five or six years old. That the poor household might have one mouth less to feed, I had been placed in grandmother's care, as I have just been saying. Here, in solitude, my first gleams of intelligence were awakened amidst the geese, the calves and the sheep. Everything before that is impenetrable darkness. My real birth is at that moment when the dawn of personality rises, dispersing the mists of unconsciousness and leaving a lasting memory. I can see myself plainly, clad in a soiled frieze frock flapping against my bare heels; I remember the handkerchief hanging from my waist by a bit of string, a handkerchief often lost and replaced by the back of my sleeve.

There I stand one day, a pensive urchin, with my hands behind my back and my face turned to the sun. The dazzling splendor fascinates me. I am the Moth attracted by the light of the lamp. With what am I enjoying the glorious radiance: with my mouth or my eyes? That is the question put by my budding scientific curiosity. Reader, do not smile: the future observer is already practicing and experimenting. I open my mouth wide and close my eyes: the glory disappears. I open my eyes and shut my mouth: the glory reappears. I repeat the performance, with the same result. The question's solved: I have learnt by deduction that I see the sun with my eyes. Oh, what a discovery! That evening, I told the whole house all about it. Grandmother smiled fondly at my simplicity: the others laughed at it. 'Tis the way of the world.

Another find. At nightfall, amidst the neighboring bushes, a sort of jingle attracted my attention, sounding very faintly and softly through the evening silence. Who is making that noise? Is it a little bird chirping in his nest? We must look into the matter and that quickly. True, there is the wolf, who comes out of the woods at this time, so they tell me. Let's go all the same, but not too far: just there, behind that clump of groom. I stand on the look out for long, but all in vain. At the faintest sound of movement in the brushwood, the jingle ceases. I try again next day and the day after. This time, my stubborn watch succeeds. Whoosh! A grab of my hand and I hold the singer. It is not a bird; it is a kind of Grasshopper whose hind legs my playfellows have taught me to like: a poor recompense for my prolonged ambush. The best part of the business is not the two haunches with the shrimpy flavor, but what I have just learnt. I now know, from personal observation, that the Grasshopper sings. I did not publish my discovery, for fear of the same laughter that greeted my story about the sun.

Oh, what pretty flowers, in a field close to the house! They seem to smile to me with their great violet eyes. Later on, I see, in their place, bunches of big red cherries. I taste them. They are not nice and they have no stones. What can those cherries be? At the end of the summer, grandfather comes with a spade and turns my field of observation topsy-turvy. From under ground there comes, by the basketful and sackful, a sort of round root. I know that root; it abounds in the house; time after time I have cooked it in the peat stove. It is the potato. Its violet flower and its red fruit are pigeonholed for good and all in my memory.

With an ever watchful eye for animals and plants, the future observer, the little six-year-old monkey, practiced by himself, all unawares. He went to the flower, he went to the insect, even as the large white butterfly goes to the cabbage and the red admiral to the thistle. He looked and inquired, drawn by a curiosity whereof heredity did not know the secret. He bore within him the germ of a faculty unknown to his family; he kept alive a glimmer that was foreign to the ancestral hearth. What will become of that infinitesimal spark of childish fancy? It will die out, beyond a doubt, unless education intervene, giving it the fuel of example, fanning it with the breath of experience. In that case, schooling will explain what heredity leaves unexplained. This is what we will examine in the next chapter.

CHAPTER VI. MY SCHOOLING

I am back in the village, in my father's house. I am now seven years old; and it is high time that I went to school. Nothing could have turned out better: the master is my godfather. What shall I call the room in which I was to become acquainted with the alphabet? It would be difficult to find the exact word, because the room served for every purpose. It was at once a school, a kitchen, a bedroom, a dining room and, at times, a chicken house and a piggery. Palatial schools were not dreamt of in those days; any wretched hovel was thought good enough.

A broad fixed ladder led to the floor above. Under the ladder stood a big bed in a boarded recess. What was there upstairs? I never quite knew. I would see the master sometimes bring down an armful of hay for the ass, sometimes a basket of potatoes which the housewife emptied into the pot in which the little porkers' food was cooked. It must have been a loft of sorts, a storehouse of provisions for man and beast. Those two apartments composed the whole building.

To return to the lower one, the schoolroom: a window faces south, the only window in the house, a low, narrow window whose frame you can touch at the same time with your head and both your shoulders. This sunny aperture is the only lively spot in the dwelling, it overlooks the greater part of the village, which straggles along the slopes of a slanting valley. In the window recess is the master's little table.

The opposite wall contains a niche in which stands a gleaming copper pail full of water. Here the parched children can relieve their thirst when they please, with a cup left within their reach. At the top of the niche are a few shelves bright with pewter plates, dishes and drinking vessels, which are taken down from their sanctuary on great occasions only.

More or less everywhere, at any spot which the light touches, are crudely colored pictures, pasted on the walls. Here is Our Lady of the Seven Dolours, the disconsolate Mother of God opening her blue cloak to show her heart pierced with seven daggers. Between the sun and moon, which stare at you with their great, round eyes, is the Eternal Father, whose robe swells as though puffed out with the storm. To the right of the window, in the embrasure, is the Wandering Jew. He wears a three-cornered hat, a large, white leather apron, hobnailed shoes and a stout stick. 'Never was such a bearded man seen before or after,' says the legend that surrounds the picture. The draftsman has not forgotten this detail: the old man's beard spreads in a snowy avalanche over the apron and comes down to his knees.

On the left is Genevieve of Brabant, accompanied by the roe, with fierce Golo hiding in the bushes, sword in hand. Above hangs The Death of Mr. Credit, slain by defaulters at the door of his inn; and so on and so on, in every variety of subject, at all the unoccupied spots of the four walls.

I was filled with admiration of this picture gallery, which held one's eyes with its great patches of red, blue, green and yellow. The master, however, had not set up his collection with a view to training our minds and hearts. That was the last and least of the worthy man's ambitions. An artist in his fashion, he had adorned his house according to his taste; and we benefited by the scheme of decoration.

While the gallery of halfpenny pictures made me happy all the year round, there was another entertainment which I found particularly attractive in winter, in frosty weather, when the snow lay long on the ground. Against the far wall stands the fireplace, as monumental in size as at my grandmother's. Its arched cornice occupies the whole width of the room, for the enormous redoubt fulfils more than one purpose. In the middle is the hearth, but, on the right and left, are two breast-high recesses, half wood and half stone. Each of them is a bed, with a mattress stuffed with chaff of winnowed corn. Two sliding planks serve as shutters and close the chest if the sleeper would be alone. This dormitory, sheltered under the chimney mantel, supplies couches for the favored ones of the house, the two boarders. They must lie snug in there at night, with their shutters closed, when the north wind howls at the mouth of the dark valley and sends the snow awhirl. The rest is occupied by the hearth and its accessories: the three-legged stools; the salt box, hanging against the wall to keep its contents dry; the heavy shovel which it takes two hands to wield; lastly, the bellows similar to those with which I used to blow out my cheeks in grandfather's house. They consist of a mighty branch of pine, hollowed throughout its length with a red-hot iron. By means of this channel, one's breath is applied, from a convenient distance, to the spot which is to be revived. With a couple of stones for supports, the master's bundle of sticks and our own logs blaze and flicker, each of us having to bring a log of wood in the morning, if he would share in the treat.

For that matter, the fire was not exactly lit for us, but, above all, to warm a row of three pots in which simmered the pigs' food, a mixture of potatoes and bran. That, despite the tribute of a log, was the real object of the brushwood fire. The two boarders, on their stools, in the best places, and we others sitting on our heels formed a semicircle around those big cauldrons, full to the brim and giving off little jets of steam, with puff-puff-puffing sounds. The bolder among us, when the master's eyes were engaged elsewhere, would dig a knife into a well cooked potato and add it to their bit of bread; for I must say that, if we did little work in my school, at least we did a deal of eating. It was the regular custom to crack a few nuts and nibble at a crust while writing our page or setting out our rows of figures.

We, the smaller ones, in addition to the comfort of studying with our mouths full, had every now and then two other delights, which were quite as good as cracking nuts. The back door communicated with the yard where the hen, surrounded by her brood of chicks, scratched at the dung hill, while the little porkers, of whom there were a dozen, wallowed in their stone trough. This door would open sometimes to let one of us out, a privilege which we abused, for the sly ones among us were careful not to close it on returning. Forthwith, the porkers would come running in, one after the other, attracted by the smell of the boiled potatoes. My bench, the one where the youngsters sat, stood against the wall, under the copper pail to which we used to go for water when the nuts had made us thirsty, and was right in the way of the pigs. Up they came trotting and grunting, curling their little tails; they rubbed against our legs; they poked their cold pink snouts into our hands in search of a scrap of crust; they questioned us with their sharp little eyes to learn if we happened to have a dry chestnut for them in our pockets. When they had gone the round, some this way and some that, they went back to the farmyard, driven away by a friendly flick of the master's handkerchief. Next came the visit of the hen, bringing her velvet-coated chicks to see us. All of us eagerly crumbled a little bread for our pretty visitors. We vied with one another in calling them to us and tickling with our fingers their soft and downy backs. No, there was certainly no lack of distractions.

What could we learn in such a school as that! Let us first speak of the young ones, of whom I was one. Each of us had, or rather was supposed to have, in his hands a little penny book, the alphabet, printed on gray paper. It began, on the cover, with a pigeon, or something like it. Next came a cross, followed by the letters in their order. When we turned over, our eyes encountered the terrible ba, be, bi, bo, bu, the stumbling block of most of us. When we had mastered that formidable page, we were considered to know how to read and were admitted among the big ones. But, if the little book was to be of any use, the least that was required was that the master should interest himself in us to some extent and show us how to set about things. For this, the worthy man, too much taken up with the big ones, had not the time. The famous alphabet with the pigeon was thrust upon us only to give us the air of scholars. We were to contemplate it on our bench, to decipher it with the help of our next neighbor, in case he might know one or two of the letters. Our contemplation came to nothing, being every moment disturbed by a visit to the potatoes in the stew pots, a quarrel among playmates about a marble, the grunting invasion of the porkers or the arrival of the chicks. With the aid of these distractions, we would wait patiently until it was time for us to go home. That was our most serious work.

The big ones used to write. They had the benefit of the small amount of light in the room, by the narrow window where the Wandering Jew and ruthless Golo faced each other, and of the large and only table with its circle of seats. The school supplied nothing, not even a drop of ink; every one had to come with a full set of utensils. The inkhorn of those days, a relic of the ancient pen case of which Rabelais speaks, was a long cardboard box divided into two stages. The upper compartment held the pens, made of goose or turkey quills trimmed with a penknife; the lower contained, in a tiny well, ink made of soot mixed with vinegar.

The master's great business was to mend the pens—a delicate work, not without danger for inexperienced fingers—and then to trace at the head of the white page a line of strokes, single letters or words, according to the scholar's capabilities. When that is over, keep an eye on the work of art which is coming to adorn the copy! With what undulating movements of the wrist does the hand, resting on the little finger, prepare and plan its flight! All at once, the hand starts off, flies, whirls; and, lo and behold, under the line of writing is unfurled a garland of circles, spirals and flourishes, framing a bird with outspread wings, the whole, if you please, in red ink, the only kind worthy of such a pen. Large and small, we stood awestruck in the presence of these marvels. The family, in the evening, after supper, would pass from hand to hand the masterpiece brought back from school: 'What a man!' was the comment. 'What a man, to draw you a Holy Ghost with a stroke of the pen!'

What was read at my school? At most, in French, a few selections from sacred history. Latin recurred oftener, to teach us to sing vespers properly. The more advanced pupils tried to decipher manuscript, a deed of sale, the hieroglyphics of some scrivener.

And history, geography? No one ever heard of them. What difference did it make to us whether the earth was round or square! In either case, it was just as hard to make it bring forth anything.

And grammar? The master troubled his head very little about that; and we still less. We should have been greatly surprised by the novelty and the forbidding look of such words in the grammatical jargon as substantive, indicative and subjunctive. Accuracy of language, whether of speech or writing, must be learnt by practice. And none of us was troubled by scruples in this respect. What was the use of all these subtleties, when, on coming out of school, a lad simply went back to his flock of sheep!

And arithmetic? Yes, we did a little of this but not under that learned name. We called it sums. To put down rows of figures, not too long, add them and subtract them one from the other was more or less familiar work. On Saturday evenings, to finish up the week, there was a general orgy of sums. The top boy stood up and, in a loud voice, recited the multiplication table up to twelve times. I say twelve times, for in those days, because of our old duodecimal measures, it was the custom to count as far as the twelve times table, instead of the ten times of the metric system. When this recital was over, the whole class, the little ones included, took it up in chorus, creating such an uproar that chicks and porkers took to flight if they happened to be there. And this went on to twelve times twelve, the first in the row starting the next table and the whole class repeating it as loud as it could yell. Of all that we were taught in school, the multiplication table was what we knew best, for this noisy method ended by dinning the different numbers into our ears. This does not mean that we became skilful reckoners. The cleverest of us easily got muddled with the figures to be carried in a multiplication sum. As for division, rare indeed were they who reached such heights. In short, the moment a problem, however insignificant, had to be solved, we had recourse to mental gymnastics much rather than to the learned aid of arithmetic.

When all is said, our master was an excellent man who could have kept school very well but for his lack of one thing; and that was time. He devoted to us all the little leisure which his numerous functions left him. And, first of all, he managed the property of an absentee landowner, who only occasionally set foot in the village. He had under his care an old castle with four towers, which had become so many pigeon houses; he directed the getting in of the hay, the walnuts, the apples and the oats. We used to help him during the summer, when the school, which was well attended in winter, was almost deserted. All that remained, because they were not yet big enough to work in the fields, were a few children, including him who was one day to set down these memorable facts. Lessons at that time were less dull. They were often given on the hay or on the straw; oftener still, lesson time was spent in cleaning out the dovecote or stamping on the snails that had sallied in rainy weather from their fortresses, the tall box borders of the garden belonging to the castle.

Our master was a barber. With his light hand, which was so clever at beautifying our copies with curlicue birds, he shaved the notabilities of the place: the mayor, the parish priest, the notary. Our master was a bell ringer. A wedding or a christening interrupted the lessons: he had to ring a peal. A gathering storm gave us a holiday: the great bell must be tolled to ward off the lightning and the hail. Our master was a choir singer. With his mighty voice, he filled the church when he led the Magnificat at vespers. Our master wound up and regulated the village clock. This was his proudest function. Giving a glance at the sun, to ascertain the time more or less nearly, he would climb to the top of the steeple, open a huge cage of rafters and find himself in a maze of wheels and springs whereof the secret was known to him alone.

With such a school and such a master and such examples, what will become of my embryo tastes, as yet so imperceptible? In that environment, they seem bound to perish, stifled for ever. Yet no, the germ has life; it works in my veins, never to leave them again. It finds nourishment everywhere, down to the cover of my penny alphabet, embellished with a crude picture of a pigeon which I study and contemplate much more zealously than the A B C. Its round eye, with its circlet of dots, seems to smile upon me. Its wing, of which I count the feathers one by one, tells me of flights on high, among the beautiful clouds; it carries me to the beeches raising their smooth trunks above a mossy carpet studded with white mushrooms that look like eggs dropped by some vagrant hen; it takes me to the snow-clad peaks where the birds leave the starry print of their red feet. He is a fine fellow, my pigeon friend: he consoles me for the woes hidden behind the cover of my book. Thanks to him, I sit quietly on my bench and wait more or less till school is over.

School out of doors has other charms. When the master takes us to kill the snails in the box borders, I do not always scrupulously fulfil my office as an exterminator. My heel sometimes hesitates before coming down upon the handful which I have gathered. They are so pretty! Just think, there are yellow ones and pink, white ones and brown, all with dark spiral streaks. I fill my pockets with the handsomest, so as to feast my eyes on them at my leisure.

On hay making days in the master's field, I strike up an acquaintance with the frog. Flayed and stuck at the end of a split stick, he serves as bait to tempt the crayfish to come out of his retreat by the brook side. On the alder trees I catch the Hoplia, the splendid scarab who pales the azure of the heavens. I pick the narcissus and learn to gather, with the tip of my tongue, the tiny drop of honey that lies right at the bottom of the cleft corolla. I also learn that too long indulgence in this feast brings a headache; but this discomfort in no way impairs my admiration for the glorious white flower, which wears a narrow red collar at the throat of its funnel.

When we go to beat the walnut trees, the barren grass plots provide me with locusts spreading their wings, some into a blue fan, others into a red. And thus the rustic school, even in the heart of winter, furnished continuous food for my interest in things. There was no need for precept and example: my passion for animals and plants made progress of itself.

What did not make progress was my acquaintance with my letters, greatly neglected in favor of the pigeon. I was still at the same stage, hopelessly behindhand with the intractable alphabet, when my father, by a chance inspiration, brought me home from the town what was destined to give me a start along the road of reading. Despite the not insignificant part which it played in my intellectual awakening, the purchase was by no means a ruinous one. It was a large print, price six farthings, colored and divided into compartments in which animals of all sorts taught the A B C by means of the first letters of their names.

Where should I keep the precious picture? As it happened, in the room set apart for the children at home, there was a little window like the one in the school, opening in the same way out of a sort of recess and in the same way overlooking most of the village. One was on the right, the other on the left of the castle with the pigeon house towers; both afforded an equally good view of the heights of the slanting valley. I was able to enjoy the school window only at rare intervals, when the master left his little table; the other was at my disposal as often as I liked. I spent long hours there, sitting on a little fixed window seat.

The view was magnificent. I could see the ends of the earth, that is to say, the hills that blocked the horizon, all but a misty gap through which the brook with the crayfish flowed under the alders and willows. High up on the skyline, a few wind-battered oaks bristled on the ridges; and beyond there lay nothing but the unknown, laden with mystery.

At the back of the hollow stood the church, with its three steeples and its clock; and, a little higher, the village square, where a spring, fashioned into a fountain, gurgled from one basin into another, under a wide arched roof. I could hear from my window the chatter of the women washing their clothes, the strokes of their beaters, the rasping of the pots scoured with sand and vinegar. Sprinkled over the slopes are little houses with their garden patches in terraces banked up by tottering walls, which bulge under the thrust of the earth. Here and there are very steep lanes, with the dents of the rock forming a natural pavement. The mule, sure-footed though he be, would hesitate to enter these dangerous passes with his load of branches.

Further on, beyond the village, half-way up the hills, stood the great ever-so-old lime tree, the Tel, as we used to call it, whose sides, hollowed out by the ages, were the favorite hiding places of us children at play. On fair days, its immense, spreading foliage cast a wide shadow over the herds of oxen and sheep. Those solemn days, which only came once a year, brought me a few ideas from without: I learnt that the world did not end with my amphitheater of hills. I saw the inn keeper's wine arrive on mule back and in goat skin bottles. I hung about the market place and watched the opening of jars full of stewed pears, the setting out of baskets of grapes, an almost unknown fruit, the object of eager covetousness. I stood and gazed in admiration at the roulette board on which, for a sou, according to the spot at which its needle stopped on a circular row of nails, you won a pink poodle made of barley sugar, or a round jar of aniseed sweets, or, much oftener, nothing at all. On a piece of canvas on the ground, rolls of printed calico with red flowers, were displayed to tempt the girls. Close by rose a pile of beechwood clogs, tops and boxwood flutes. Here the shepherds chose their instruments, trying them by blowing a note or two. How new it all was to me! What a lot of things there were to see in this world! Alas, that wonderful time was of but short duration! At night, after a little brawling at the inn, it was all over; and the village returned to silence for a year.

But I must not linger over these memories of the dawn of life. We were speaking of the memorable picture brought from town. Where shall I keep it, to make the best use of it? Why, of course, it must be pasted on the embrasure of my window. The recess, with its seat, shall be my study cell; here I can feast my eyes by turns on the big lime tree and the animals of my alphabet. And this was what I did.

And now, my precious picture, it is our turn, yours and mine. You began with the sacred beast, the ass, whose name, with a big initial, taught me the letter A. The boeuf, the ox, stood for B; the canard, the duck, told me about C; the dindon, the turkey, gave me the letter D. And so on with the rest. A few compartments, it is true, were lacking in clearness. I had no friendly feeling for the hippopotamus, the kamichi, or horned screamer, and the zebu, who aimed at making me say H, K and Z. Those outlandish beasts, which failed to give the abstract letter the support of a recognized reality, caused me to hesitate for a time over their recalcitrant consonants. No matter: father came to my aid in difficult cases; and I made such rapid progress that, in a few days, I was able to turn in good earnest the pages of my little pigeon book, hitherto so undecipherable. I was initiated; I knew how to spell. My parents marveled. I can explain this unexpected progress today. Those speaking pictures, which brought me amongst my friends the beasts, were in harmony with my instincts. If the animal has not fulfilled all that it promised in so far as I am concerned, I have at least to thank it for teaching me to read. I should have succeeded by other means, I do not doubt, but not so quickly nor so pleasantly. Animals forever!

Luck favored me a second time. As a reward for my prowess, I was given La Fontaine's Fables, in a popular, cheap edition, crammed with pictures, small, I admit, and very inaccurate, but still delightful. Here were the crow, the fox, the wolf, the magpie, the frog, the rabbit, the ass, the dog, the cat: all persons of my acquaintance. The glorious book was immensely to my taste, with its skimpy illustrations on which the animal walked and talked. As to understanding what it said, that was another story! Never mind, my lad! Put together syllables that say nothing to you as yet; they will speak to you later and La Fontaine will always remain your friend.

I come to the time when I was ten years old and at Rodez College. My functions as a serving boy in the chapel entitled me to free instruction as a day boarder. There were four of us in white surplices and red skull-caps and cassocks. I was the youngest of the party and did little more than walk on. I counted as a unit; and that was about all, for I was never certain when to ring the bell or move the missal. I was all of a tremble when we gathered two on this side and two on that, with genuflection's, in the middle of the sanctuary, to intone the Domine, salvum fac regern at the end of mass. Let me make a confession: tongue-tied with shyness, I used to leave it to the others.

Nevertheless, I was well thought of, for, in the school, I cut a good figure in composition and translation. In that classical atmosphere, there was talk of Procas, King of Alba, and of his two sons, Numitor and Amulius. We heard of Cynoegirus, the strong jawed man, who, having lost his two hands in battle, seized and held a Persian galley with his teeth, and of Cadmus the Phoenician, who sowed a dragon's teeth as though they were beans and gathered his harvest in the shape of a host of armed men, who killed one another as they rose up from the ground. The only one who survived the slaughter was one as tough as leather, presumably the son of the big back grinder.

Had they talked to me about the man in the moon, I could not have been more startled. I made up for it with my animals, which I was far from forgetting amid this phantasmagoria of heroes and demigods. While honoring the exploits of Cadmus and Cynoegirus, I hardly ever failed, on Sundays and Thursdays [the weekly half-holiday in French schools], to go and see if the cowslip or the yellow daffodil was making its appearance in the meadows, if the Linnet was hatching on the juniper bushes, if the Cockchafers were plopping down from the wind shaken poplars. Thus was the sacred spark kept aglow, ever brighter than before.

By easy stages, I came to Virgil and was very much smitten with Meliboeus, Corydon, Menalcas, Damoetas and the rest of them. The scandals of the ancient shepherds fortunately passed unnoticed; and within the frame in which the characters moved were exquisite details concerning the bee, the cicada, the turtle dove, the crow, the nanny goat and the golden broom. A veritable delight were these stories of the fields, sung in sonorous verse; and the Latin poet left a lasting impression on my classical recollections.

Then, suddenly, goodbye to my studies, goodbye to Tityrus and Menalcas. Ill luck is swooping down on us, relentlessly. Hunger threatens us at home. And now, boy, put your trust in God; run about and earn your penn'orth of potatoes as best you can. Life is about to become a hideous inferno. Let us pass quickly over this phase. Amid this lamentable chaos, my love for the insect ought to have gone under. Not at all. It would have survived the raft of the Medusa. I still remember a certain pine cockchafer met for the first time. The plumes on her antennae, her pretty pattern of white spots on a dark brown ground were as a ray of sunshine in the gloomy wretchedness of the day.

To cut a long story short: good fortune, which never abandons the brave, brought me to the primary normal school at Vaucluse where I was assured food: dried chestnuts and chickpeas. The principal, a man of broad views, soon came to trust his new assistant. He left me practically a free hand, so long as I satisfied the school curriculum, which was very modest in those days. Possessing a smattering of Latin and grammar, I was a little ahead of my fellow pupils. I took advantage of this to get some order into my vague knowledge of plants and animals. While a dictation lesson was being corrected around me, with generous assistance from the dictionary, I would examine, in the recesses of my desk, the oleander's fruit, the snapdragon's seed vessel, the wasp's sting and the ground beetle's wing-case.

With this foretaste of natural science, picked up haphazard and by stealth, I left school more deeply in love than ever with insects and flowers. And yet I had to give it all up. That wider education, which would have to be my source of livelihood in the future, demanded this imperiously. What was I to take in hand to raise me above the primary school, whose staff could barely earn their bread in those days? Natural history could not bring me anywhere. The educational system of the time kept it at a distance, as unworthy of association with Latin and Greek. Mathematics remained, with its very simple equipment: a blackboard, a bit of chalk and a few books.

So I flung myself with might and main into conic sections and the calculus: a hard battle, if ever there was one, without guides or counselors, face to face for days on end with the abstruse problem which my stubborn thinking at last stripped of its mysteries. Next came the physical sciences, studied in the same manner, with an impossible laboratory, the work of my own hands.

The reader can imagine the fate of my favorite branch of science in this fierce struggle. At the faintest sign of revolt, I lectured myself severely, lest I should let myself be seduced by some new grass, some unknown Beetle. I did violence to my feelings. My natural history books were sentenced to oblivion, relegated to the bottom of a trunk.

And so, in the end, I am sent to teach physics and chemistry at Ajaccio College. This time, the temptation is too much for me. The sea, with its wonders, the beach, whereon the tide casts such beautiful shells, the maquis of myrtles, arbutus and mastic trees: all this paradise of gorgeous nature has too much on its side in the struggle with the sine and the cosine. I succumb. My leisure time is divided into two parts. One, the larger, is allotted to mathematics, the foundation of my academical future, as planned by myself; the other is spent, with much misgiving, in botanizing and looking for the treasures of the sea. What a country and what magnificent studies to be made, if, unobsessed by x and y, I had devoted myself wholeheartedly to my inclinations!

We are the wisp of straw, the plaything of the winds. We think that we are making for a goal deliberately chosen; destiny drives us towards another. Mathematics, the exaggerated preoccupation of my youth, did me hardly any service; and animals, which I avoided as much as ever I could, are the consolation of my old age. Nevertheless, I bear no grudge against the sine and the cosine, which I continue to hold in high esteem. They cost me many a pallid hour at one time, but they always afforded me some first rate entertainment: they still do so, when my head lies tossing sleeplessly on its pillow.

Meanwhile, Ajaccio received the visit of a famous Avignon botanist, Requien by name, who, with a box crammed with paper under his arm, had long been botanizing all over Corsica, pressing and drying specimens and distributing them to his friends. We soon became acquainted. I accompanied him in my free time on his explorations and never did the master have a more attentive disciple. To tell the truth, Requien was not a man of learning so much as an enthusiastic collector. Very few would have felt capable of competing with him when it came to giving the name or the geographical distribution of a plant. A blade of grass, a pad of moss, a scab of lichen, a thread of seaweed: he knew them all. The scientific name flashed across his mind at once. What an unerring memory, what a genius for classification amid the enormous mass of things observed! I stood aghast at it. I owe much to Requien in the domain of botany. Had death spared him longer, I should doubtless have owed more to him, for his was a generous heart, ever open to the troubles of novices.

In the following year, I met Moquin-Tandon, with whom, thanks to Requien, I had already exchanged a few letters on botany. The illustrious Toulouse professor came to study on the spot the flora which he proposed to describe systematically. When he arrived, all the hotel bedrooms were reserved for the members of the general council which had been summoned; and I offered him board and lodging: a shakedown in a room overlooking the sea; fare consisting of lampreys, turbot and sea urchins: common enough dishes in that land of Cockayne, but possessing no small attraction for the naturalist, because of their novelty. My cordial proposal tempted him; he yielded to my blandishments; and there we were for a fortnight chatting at table de omni re scibili after the botanical excursion was over.

With Moquin-Tandon, new vistas opened before me. Here it was no longer the case of a nomenclator with an infallible memory: he was a naturalist with far-reaching ideas, a philosopher who soared above petty details to comprehensive views of life, a writer, a poet who knew how to clothe the naked truth in the magic mantle of the glowing word. Never again shall I sit at an intellectual feast like that: 'Leave your mathematics,' he said. 'No one will take the least interest in your formula. Get to the beast, the plant; and, if, as I believe, the fever burns in your veins, you will find men to listen to you.'

We made an expedition to the center of the island, to Monte Renoso, with which I was already familiar. I made the scientist pick the hoary everlasting (Helichrysum frigidum), which makes a wonderful patch of silver; the many-headed thrift, or mouflon grass (Armeria multiceps), which the Corsicans call erba muorone; the downy marguerite (Leucanthemum tomosum), which, clad in wadding, shivers amid the snows; and many other rarities dear to the botanist. Moquin-Tandon was jubilant. I, on my side, was much more attracted and overcome by his words and his enthusiasm than by the hoary everlasting. When we came down from the cold mountaintop, my mind was made up: mathematics would be abandoned.

On the day before his departure, he said to me: 'You interest yourself in shells. That is something, but it is not enough. You must look into the animal itself. I will show you how it's done.'

And, taking a sharp pair of scissors from the family work-basket and a couple of needles stuck into a bit of vine shoot which served as a makeshift handle, he showed me the anatomy of a snail in a soup plate filled with water. Gradually he explained and sketched the organs which he spread before my eyes. This was the only, never-to-be-forgotten lesson in natural history that I ever received in my life.

It is time to conclude. I was cross-examining myself, being unable to cross-examine the silent Beetle. As far as it is possible to read within myself, I answer as follows: 'From early childhood, from the moment of my first mental awakening, I have felt drawn towards the things of nature, or, to return to our catchword, I have the gift, the bump of observation.'

After the details which I have already given about my ancestors, it would be ridiculous to look to heredity for an explanation of the fact. Nor would any one venture to suggest the words or example of my masters. Of scientific education, the fruit of college training, I had none whatever. I never set foot in a lecture hall except to undergo the ordeal of examinations. Without masters, without guides, often without books, in spite of poverty, that terrible extinguisher, I went ahead, persisted, facing my difficulties, until the indomitable bump ended by shedding its scanty contents. Yes, they were very scanty, yet possibly of some value, if circumstances had come to their assistance. I was a born animalist. Why and how? No reply.

We thus have, all of us, in different directions and in a greater or lesser degree, characteristics that brand us with a special mark, characteristics of an unfathomable origin. They exist because they exist; and that is all that any one can say. The gift is not handed down: the man of talent has a fool for a son. Nor is it acquired; but it is improved by practice. He who has not the germ of it in his veins will never possess it, in spite of all the pains of a hothouse education.

That to which we give the name of instinct when speaking of animals is something similar to genius. It is, in both cases, a peak that rises above the ordinary level. But instinct is handed down, unchanged and undiminished, throughout the sequence of a species; it is permanent and general and in this it differs greatly from genius, which is not transmissible and changes in different cases. Instinct is the inviolable heritage of the family and falls to one and all, without distinction. Here the difference ends. Independent of similarity of structure, it breaks out like genius, here or elsewhere, for no perceptible reason. Nothing causes it to be foreseen, nothing in the organization explains it. If cross-examined on this point, the Dung beetles and the rest, each with his own peculiar talent, would answer, were we able to understand them: 'Instinct is the animal's genius.'

CHAPTER VII. THE POND

The pond, the delight of my early childhood, is still a sight whereof my old eyes never tire. What animation in that verdant world! On the warm mud of the edges, the frog's little tadpole basks and frisks in its black legions; down in the water, the orange-bellied newt steers his way slowly with the broad rudder of his flat tail; among the reeds are stationed the flotillas of the caddis worms, half protruding from their tubes, which are now a tiny bit of stick and again a turret of little shells.

In the deep places, the water beetle dives, carrying with him his reserves of breath: an air bubble at the tip of the wing cases and, under the chest, a film of gas that gleams like a silver breastplate; on the surface, the ballet of those shimmering pearls, the whirligigs, turns and twists about; hard by there skims the unsubmersible troop of the pond skaters, who glide along with side strokes similar to those which the cobbler makes when sewing.

Here are the water boatmen, who swim on their backs with two oars spread cross-wise, and the flat water scorpions; here, squalidly clad in mud, is the grub of the largest of our dragonflies, so curious because of its manner of progression: it fills its hinder parts, a yawning funnel, with water, spurts it out again and advances just so far as the recoil of its hydraulic cannon.

The mollusks abound, a peaceful tribe. At the bottom, the plump river snails discreetly raise their lid, opening ever so little the shutters of their dwelling; on the level of the water, in the glades of the aquatic garden, the pond snails—Physa, Limnaea and Planorbis—take the air. Dark leeches writhe upon their prey, a chunk of earthworm; thousands of tiny, reddish grubs, future mosquitoes, go spinning around and twist and curve like so many graceful dolphins.

Yes, a stagnant pool, though but a few feet wide, hatched by the sun, is an immense world, an inexhaustible mine of observation to the studious man and a marvel to the child who, tired of his paper boat, diverts his eyes and thoughts a little with what is happening in the water. Let me tell what I remember of my first pond, at a time when ideas began to dawn in my seven-year-old brain.

How shall a man earn his living in my poor native village, with its inclement weather and its niggardly soil? The owner of a few acres of grazing land rears sheep. In the best parts, he scrapes the soil with the swing plow; he flattens it into terraces banked by walls of broken stones. Pannierfuls of dung are carried up on donkey-back from the cowshed.

Then, in due season, comes the excellent potato, which, boiled and served hot in a basket of plaited straw, is the chief stand-by in winter.

Should the crop exceed the needs of the household, the surplus goes to feed a pig, that precious beast, a treasure of bacon and ham. The ewes supply butter and curds; the garden boasts cabbages, turnips and even a few hives in a sheltered corner. With wealth like that one can look fate in the face.

But we, we have nothing, nothing but the little house inherited by my mother and its adjoining patch of garden. The meager resources of the family are coming to an end. It is time to see to it and that quickly. What is to be done? That is the stern question which father and mother sat debating one evening.

Hop-o'-my-Thumb, hiding under the woodcutter's stool, listened to his parents overcome by want. I also, pretending to sleep, with my elbows on the table, listen not to blood curdling designs, but to grand plans that set my heart rejoicing. This is how the matter stands: at the bottom of the village, near the church, at the spot where the water of the large roofed spring escapes from its underground weir and joins the brook in the valley, an enterprising man, back from the war, has set up a small tallow factory. He sells the scrapings of his pans, the burnt fat, reeking of candle grease, at a low price. He proclaims these wares to be excellent for fattening ducks.

"Suppose we bred some ducks," says mother. "They sell very well in town. Henri would mind them and take them down to the brook."

"Very well," says father, "let's breed some ducks. There may be difficulties in the way; but we'll have a try."

That night, I had dreams of paradise: I was with my ducklings, clad in their yellow suits; I took them to the pond, I watched them have their bath, I brought them back again, carrying the more tired ones in a basket.

A month or two after, the little birds of my dreams were a reality. There were twenty-four of them. They had been hatched by two hens, of whom one, the big, black one, was an inmate of the house, while the other was borrowed from a neighbor.

To bring them up, the former is sufficient, so careful is she of her adopted family. At first, everything goes perfectly: a tub with two fingers' depth of water serves as a pond. On sunny days, the ducklings bathe in it under the anxious eye of the hen.

A fortnight later, the tub is no longer enough. It contains neither cresses crammed with tiny shellfish nor worms and tadpoles, dainty morsels both. The time has come for dives and hunts amid the tangle of the water weeds;

and for us the day of trouble has also come. True, the miller, down by the brook, has fine ducks, easy and cheap to bring up; the tallow smelter, who has extolled his burnt fat so loudly, has some as well, for he has the advantage of the waste water from the spring at the bottom of the village; but how are we, right up there, at the top, to procure aquatic sports for our broods? In summer, we have hardly water to drink!

Near the house, in a freestone recess, a scanty source trickles into a basin made in the rock.. Four or five families have, like ourselves, to draw their water there with copper pails. By the time that the schoolmaster's donkey has slaked her thirst and the neighbors have taken their provision for the day, the basin is dry. We have to wait for four-and-twenty hours for it to fill. No, this is not the hole in which the ducks would delight nor indeed in which they would be tolerated.

There remains the brook. To go down to it with the troop of ducklings is fraught with danger. On the way through the village, we might meet cats, bold ravishers of small poultry; some surly mongrel might frighten and scatter the little band; and it would be a hard puzzle to collect it in its entirety. We must avoid the traffic and take refuge in peaceful and sequestered spots.

On the hills, the path that climbs behind the chateau soon takes a sudden turn and widens into a small plain beside the meadows. It skirts a rocky slope whence trickles, level with the ground, a streamlet, forming a pond of some size. Here profound solitude reigns all day long. The ducklings will be well off; and the journey can be made in peace by a deserted footpath.

You, little man, shall take them to that delectable spot. What a day it was that marked my first appearance as a herdsman of ducks! Why must there be a jar to the even tenor of such joys? The too frequent encounter of my tender skin with the hard ground had given me a large and painful blister on the heel. Had I wanted to put on the shoes stowed away in the cupboard for Sundays and holidays, I could not. There was nothing for it but to go barefoot over the broken stones, dragging my leg and carrying high the injured heel.

Let us make a start, hobbling along, switch in hand, behind the ducks. They too, poor little things, have sensitive soles to their feet; they limp, they quack with fatigue. They would refuse to go any farther if I did not, from time to time, call a halt under the shelter of an ash.

We are there at last. The place could not be better for my birdlets; shallow, tepid water, interspersed with muddy knolls and green eyots. The diversions of the bath begin forthwith. The ducklings clap their beaks and rummage here, there and everywhere; they sift each mouthful, rejecting the clear water and retaining the good bits. In the deeper parts, they point their sterns into the air and stick their heads under water. They are happy; and it is a blessed thing to see them at work. We will let them be. It is my turn to enjoy the pond.

What is this? On the mud lie some loose, knotted, soot-colored cords. One could take them for threads of wool like those which you pull out of an old ravelly stocking. Can some shepherdess, knitting a black sock and finding her work turn out badly, have begun all over again and, in her impatience, have thrown down the wool with all the dropped stitches? It really looks like it.

I take up one of those cords in my hand. It is sticky and extremely slack; the thing slips through the fingers before they can catch hold of it. A few of the knots burst and shed their contents. What comes out is a black globule, the size of a pin's head, followed by a flat tail. I recognize, on a very small scale, a familiar object: the tadpole, the frog's baby. I have seen enough. Let us leave the knotted cords alone.

The next creatures please me better. They spin round on the surface of the water and their black backs gleam in the sun. If I lift a hand to seize them, that moment they disappear, I know not where. It's a pity: I should have much liked to see them closer and to make them wriggle in a little bowl which I should have put ready for them.

Let us look at the bottom of the water, pulling aside those bunches of green string whence beads of air are rising and gathering into foam. There is something of everything underneath. I see pretty shells with compact whorls, flat as beans; I notice little worms carrying tufts and feathers; I make out some with flabby fins constantly flapping on their backs. What are they all doing there? What are their names? I do not know. And I stare at them for ever so long, held by the incomprehensible mystery of the waters.

At the place where the pond dribbles into the adjoining field are some alder trees; and here I make a glorious find. It is a scarab—not a very large one, oh no! He is smaller than a cherry-stone, but of an unutterable blue. The angels in paradise must wear dresses of that color. I put the glorious one inside an empty snail-shell, which I plug up with a leaf. I shall admire that living jewel at my leisure, when I get back. Other distractions summon me away.

The spring that feeds the pond trickles from the rock, cold and clear. The water first collects into a cup, the size of the hollow of one's two hands, and then runs over in a stream. These falls call for a mill: that goes without saying. Two bits of straw, artistically crossed upon an axis, provide the machinery; some flat stones set on edge afford supports. It is a great success: the mill turns admirably. My triumph would be complete, could I but share it. For want of other playmates, I invite the ducks.

Everything palls in this poor world of ours, even a mill made of two straws. Let us think of something else: let us contrive a dam to hold back the waters and form a pool. There is no lack of stones for the brickwork. I pick the most suitable; I break the larger ones. And, while collecting these blocks, suddenly I forget all about the dam which I meant to build.

On one of the broken stones, in a cavity large enough for me to put my fist in, something gleams like glass. The hollow is lined with facets gathered in sixes which flash and glitter in the sun. I have seen something like this in church, on the great saints' days, when the light of the candles in the big chandelier kindles the stars in its hanging crystal.

We children, lying, in summer, on the straw of the threshing floor, have told one another stories of the treasures which a dragon guards underground. Those treasures now return to my mind: the names of precious stones ring out uncertainly but gloriously in my memory. I think of the king's crown, of the princesses' necklaces. In breaking stones, can I have found, but on a much richer scale, the thing that shines quite small in my mother's ring? I want more such.

The dragon of the subterranean treasures treats me generously. He gives me his diamonds in such quantities that soon I possess a heap of broken stones sparkling with magnificent clusters. He does more: he gives me his gold. The trickle of water from the rock falls on a bed of fine sand which it swirls into bubbles. If I bent over towards the light, I see something like gold filings whirling where the fall touches the bottom. Is it really the famous metal of which twenty-franc pieces, so rare with us at home, are made? One would think so, from the glitter.

I take a pinch of sand and place it in my palm. The brilliant particles are numerous, but so small that I have to pick them up with a straw moistened in my mouth. Let us drop this: they are too tiny and too bothersome to collect. The big, valuable lumps must be farther on, in the thickness of the rock. We'll come back later; we'll blast the mountain.

I break more stones. Oh, what a queer thing has just come loose, all in one piece! It is turned spiral-wise, like certain flat snails that come out of the cracks of old walls in rainy weather. With its gnarled sides, it looks like a little ram's horn. Shell or horn, it is very curious. How do things like that find their way into the stone?

Treasures and curiosities make my pockets bulge with pebbles. It is late and the little ducklings have had all they want to eat. Come along, youngsters, let's go home. My blistered heel is forgotten in my excitement. The walk back is a delight. A voice sings in my ear, an untranslatable voice, softer than any language and bewildering as a dream. It speaks to me for the first time of the mysteries of the pond; it glorifies the heavenly insect which I hear moving in the empty snail shell, its temporary cage; it whispers the secrets of the rock, the gold filings, the faceted jewels, the ram's horn turned to stone.

Poor simpleton, smother your joy! I arrive. My parents catch sight of my bulging pockets, with their disgraceful load of stones. The cloth has given way under the rough and heavy burden.

"You rascal!" says father, at sight of the damage. "I send you to mind the ducks and you amuse yourself picking up stones, as though there weren't enough of them all round the house! Make haste and throw them away!"

Broken hearted, I obey. Diamonds, gold dust, petrified ram's horn, heavenly beetle are all flung on a rubbish heap outside the door.

Mother bewails her lot: "A nice thing, bringing up children to see them turn out so badly! You'll bring me to my grave. Green stuff I don't mind: it does for the rabbits. But stones, which ruin your pockets; poisonous animals, which'll sting your hand: what good are they to you, silly? There's no doubt about it: some one has thrown a spell over you!"

Yes, my poor mother, you were right, in your simplicity: a spell had been cast upon me; I admit it today. When it is hard enough to earn one's bit of bread, does not improving one's mind but render one more meet for suffering? Of what avail is the torment of learning to the derelicts of life?

A deal better off am I, at this late hour, dogged by poverty and knowing that the diamonds of the duck pool were rock crystal, the gold dust mica, the stone horn an Ammonite and the sky-blue beetle a Hoplia! We poor men would do better to mistrust the joys of knowledge: let us dig our furrow in the fields of the commonplace, flee the temptations of the pond, mind our ducks and leave to others, more favored by fortune, the job of explaining the world's mechanism, if the spirit moves them.

And yet no! Alone among living creatures, man has the thirst for knowledge; he alone pries into the mysteries of things. The least among us will utter his whys and his wherefores, a fine pain unknown to the brute beast. If these questionings come from us with greater persistence, with a more imperious authority, if they divert us from the quest of lucre, life's only object in the eyes of most men, does it become us to complain? Let us be careful not to do so, for that would be denying the best of all our gifts.

Let us strive, on the contrary, within the measure of our capacity, to force a gleam of light from the vast unknown; let us examine and question and, here and there, wrest a few shreds of truth. We shall sink under the task; in the present ill ordered state of society, we shall end, perhaps, in the workhouse. Let us go ahead for all that: our consolation shall be that we have increased by one atom the general mass of knowledge, the incomparable treasure of mankind.

As this modest lot has fallen to me, I will return to the pond, notwithstanding the wise admonitions and the bitter tears which I once owed to it. I will return to the pond, but not to that of the small ducks, the pond aflower with illusions: those ponds do not occur twice in a lifetime. For luck like that, you must be in all the new glory of your first breeches and your first ideas.

Many another have I come upon since that distant time, ponds very much richer and, moreover, explored with the ripened eye of experience. Enthusiastically I searched them with the net, stirred up their mud, ransacked their trailing weeds. None in my memories comes up to the first, magnified in its delights and mortifications by the marvelous perspective of the years.

Nor would any of them suit my plans of today. Their world is too vast. I should lose myself in their immensities, where life swarms freely in the sun. Like the ocean, they are infinite in their fruitfulness. And then any assiduous watching, undisturbed by passers by, is an impossibility on the public way. What I want is a pond on an extremely reduced scale, sparingly stocked in my own fashion an artificial pond standing permanently on my study table.

A louis has been overlooked in a corner of the drawer. I can spend it without seriously jeopardizing the domestic balance. Let me make this gift to science, who, I fear, will be none too much obliged to me. A gorgeous equipment may be all very well for laboratories wherein the cells and fibers of the dead are consulted at great expense; but such magnificence is of doubtful utility when we have to study the actions of the living. It is the humble makeshift, of no value, that stumbles on the secrets of life.

What did the best results of my studies of instinct cost me? Nothing but time and, above all, patience. My extravagant expenditure of twenty francs, therefore, will be a risky speculation if devoted to the purchase of an apparatus of study. It will bring me in nothing in the way of fresh views, of that I am convinced. However, let us try.

The blacksmith makes me the framework of a cage out of a few iron rods. The joiner, who is also a glazier on occasion—for, in my village, you have to be a Jack-of-all-trades if you would make both ends meet—sets the framework on a wooden base and supplies it with a movable board as a lid; he fixes thick panes of glass in the four sides. Behold the apparatus, complete, with a bottom of tarred sheet iron and a trap to let the water out.

The makers express themselves satisfied with their work, a singular novelty in their respective shops, where many an inquisitive caller has wondered what use I intend to make of my little glass trough. The thing creates a certain stir. Some insist that it is meant to hold my supplies of oil and to take the place of the receptacle in general use in our parts, the urn dug out of a block of stone. What would those utilitarians have thought of my crazy mind, had they known that my costly gear would merely serve to let me watch some wretched animals kicking about in the water!

Smith and glazier are content with their work. I myself am pleased. For all its rustic air, the apparatus does not lack elegance. It looks very well, standing on a little table in front of a window visited by the sun for the greater part of the day. Its holding capacity is some ten or eleven gallons. What shall we call it? An aquarium? No, that would be too pretentious and would, very unjustly, suggest the aquatic toy filled with rock work, waterfalls and goldfish beloved of the dwellers in suburbia. Let us preserve the gravity of serious things and not treat my learned trough as though it were a drawing room futility. We will call it the glass pond.

I furnish it with a heap of those limy incrustations wherewith certain springs in the neighborhood cover the dead clump of rushes. It is light, full of holes and gives a faint suggestion of a coral reef. Moreover, it is covered with a short, green, velvety moss, a downy sward of infinitesimal pond weed. I count on this modest vegetation to keep the water in a reasonably wholesome state, without driving me to frequent renewals which would disturb the work of my colonies. Sanitation and quiet are the first conditions of success. Now the stocked pond will not be long in filling itself with gases unfit to breathe, with putrid effluvia and other animal refuse; it will become a sink in which life will have killed life. Those dregs must disappear as soon as they are formed, must be burnt and purified; and from their oxidized ruins there must even rise a perfect life-giving gas, so that the water may retain an unchangeable store of the breathable element. The plant effects this purification in its sewage farm of green cells.

When the sun beats upon the glass pond, the work of the water weeds is a sight to behold. The green-carpeted reef is lit up with an infinity of scintillating points and assumes the appearance of a fairy lawn of velvet, studded with thousands of diamond pin's heads. From this exquisite jewelry pearls break loose continuously and are at once replaced by others in the generating casket; slowly they rise, like tiny globes of light. They spread on every side. It is a constant display of fireworks in the depths of the water.

Chemistry tells us that, thanks to its green matter and the stimulus of the sun's rays, the weeds decompose the carbonic acid gas wherewith the water is impregnated by the breathing of its inhabitants and the corruption of the organic refuse; it retains the carbon, which is wrought into fresh tissues; it exhales the oxygen in tiny bubbles. These partly dissolve in the water and partly reach the surface, where their froth supplies the atmosphere with an excess of breathable gas. The dissolved portion keeps the colonists of the pond alive and causes the unhealthy products to be oxidized and disappear.

Old hand though I be, I take an interest in this trite marvel of a bundle of weeds perpetuating hygienic principles in a stagnant pool; I look with a delighted eye upon the inexhaustible spray of spreading bubbles; I see in imagination the prehistoric times when seaweed, the first-born of plants, produced the first atmosphere for living things to breathe at the time when the silt of the continents was beginning to emerge. What I see before my eyes, between the glass panes of my trough, tells me the story of the planet surrounding itself with pure air.

CHAPTER VIII. THE CADDIS WORM

Whom shall I lodge in my glass trough, kept permanently wholesome by the action of the water weeds? I shall keep caddis worms, those expert dressers. Few of the self-clothing insects surpass them in ingenious attire. The ponds in my neighborhood supply me with five or six species, each possessing an art of its own. Today, but one of these shall receive historical honors.

I obtain it from the muddy bottomed, stagnant pools crammed with small reeds. As far as one can judge from the habitation merely, it should be, according to the specialists, Limnophilus flavicornis, whose work has earned for the whole corporation the pretty name of Phryganea, a Greek term meaning a bit of wood, a stick. In a no less expressive fashion, the Provencal peasant calls it lou portofais, lou porto-caneu. This is the little grub that carries through the still waters a faggot of tiny fragments fallen from the reeds.

Its sheath, a travelling house, is a composite and barbaric piece of work, a megalithic pile wherein art, retires in favor of amorphous strength. The materials are many and sundry, so much so that we might imagine that we had the work of dissimilar builders before our eyes, if frequent transitions did not tell us the contrary.

With the young ones, the novices, it starts with a sort of deep basket in rustic wicker-work. The twigs employed present nearly always the same characteristics and are none other than bits of small, stiff roots, long steeped and peeled under water. The grub that has made a find of these fibers saws them with its mandibles and cuts them into little straight sticks, which it fixes one by one to the edge of its basket, always crosswise, perpendicular to the axis of the work.

Picture a circle surrounded by a bristling mass of tangents, or rather a polygon with its sides extended in all directions. On this assemblage of straight lines we place repeated layers of others, without troubling about similarity of position, thus obtaining a sort of ragged fascine, whose sticks project on every side. Such is the bastion of the child grub, an excellent system of defense, with its continuous pile of spikes, but difficult to steer through the tangle of aquatic plants.

Sooner or later, the worm forsakes this kind of caltrop which catches on to everything. It was a basket maker, it now turns carpenter; it builds with little beams and joists—that is to say, with round bits of wood, browned by the water, often as wide as a thick straw and a finger's-breadth long, more or less—taking them as chance supplies them.

For the rest, there is something of everything in this rag bag: bits of stubble, fag ends of rushes, scraps of plants, fragments of some tiny twig or other, chips of wood, shreds of bark, largish grains, especially the seeds of the yellow iris, which were red when they fell from their capsules and are now black as jet.

The heterogeneous collection is piled up anyhow. Some pieces are fixed lengthwise, others across, others aslant. There are angles in this direction and angles in the other, resulting in sharp little turns and twists; the big is mixed with the little, the correct rubs shoulders with the shapeless. It is not an edifice, it is a frenzied conglomeration. Sometimes, a fine disorder is an effect of art. This is not so here: the work of the Caddis worm is not a masterpiece worth signing.

And this mad heaping up follows straight upon the regular basket work of the start. The young grub's fascine did not lack a certain elegance, with its dainty laths, all stacked crosswise, methodically; and, lo and behold, the builder, grown larger, more experienced and, one would think, more skilful, abandons the orderly plan to adopt another which is wild and incoherent! There is no transition stage between the two systems. The extravagant pile rises abruptly from the original basket. But that we often find the two kinds of work placed one above the other, we would not dare ascribe to them a common origin. The fact of their being joined together is the only thing that makes them one, in spite of the incongruity.

But the two storeys do not last indefinitely. When the worm has grown slightly and is housed to its satisfaction in a heap of joists, it abandons the basket of its childhood, which has become too narrow and is now a troublesome burden. It cuts through its sheath, lops off and lets go the stern, the original work. When moving to a higher and roomier flat, it understands how to lighten its portable house by breaking off a part of it. All that remains is the upper floor, which is enlarged at the aperture, as and when required, by the same architecture of disordered beams.

Side by side with these cases, which are mere ugly faggots, we find others just as often of exquisite beauty and composed entirely of tiny shells. Do they come from the same workshop? It takes very convincing proofs to make us believe this. Here is order with its charm, there disorder with its hideousness; on the one hand a dainty mosaic of shells, on the other a clumsy heap of sticks. And yet it is all produced by the same laborer.

Proofs abound. On some case which offends the eye with the want of arrangement in its bits of wood, patches are apt to appear which are quite regular and made of shells; in the same way, it is not unusual to see a horrid tangle of joists braced to a masterpiece of shell work. One feels a certain annoyance at seeing the pretty sheath so barbarously spoilt.

This mixed construction tells us that the rustic stacker of wooden beams excels, when occasion offers, in making elegant shell pavements and that it practices rough carpentry and delicate mosaic work indifferently. In the latter instance, the scabbard is made, above all, of Planorbes, selected among the smaller of these pond snails and laid flat. Without being scrupulously regular, the work, at its best, does not lack merit. The pretty, close-whorled spirals, placed one against the other on the same level, have a very pleasing general effect. No pilgrim returning from Santiago de Compostella ever slung handsomer tippet from his shoulders.

But only too often the caddis worm dashes ahead, regardless of proportion. The big is joined to the small, the exaggerated suddenly stands out, to the great detriment of order. Side by side with tiny Planorbes, each at most the size of a lentil, others are fixed as large as one's fingernail; and these cannot possibly be fitted in correctly. They overlap the regular parts and spoil their finish.

To crown the disorder, the caddis worm adds to the flat spirals any dead shell that comes handy, without distinction of species, provided it be not excessively large. I notice, in its collection of bric-a-brac, the Physa, the Paludina, the Limnaea, the Amber snail [all pond snails] and even the Pisidium [a bivalve], that little twin-valved casket.

Land shells, swept into the ditches by the rains after the inmate's death, are accepted quite as readily. In the work made of the Mollusk's cast-off clothing, I find encrusted the spindle shell of the Clausilium, the key shell of the pupa, the spiral of the smaller Helix, the yawning volute of the Vitrina, or glass snail, the turret shell of the Bulimus [all land snails], denizens all of the fields. In short, the caddis worm builds with more or less everything that comes from the plant or the dead mollusk. Among the diversified refuse of the pond, the only materials rejected are those of a gravelly nature. Stone and pebble are excluded from the building with a care that is very rarely absent. This is a question of hydrostatics to which we will return presently. For the moment, let us try to follow the construction of the scabbard.

In a tumbler small enough to allow of easy and precise observation, I install three or four caddis worms, extracted this moment from their sheaths with every possible precaution. After a number of attempts which have at last shown me the right road, I place at their disposal two kinds of materials, possessing opposite qualities; the supple and the firm, the soft and the hard. On the one hand, we have a live aquatic plant, such as watercress, for instance, or ombrelle d'eau, having at its base a tufty bunch of fine white roots about as thick as a horsehair. In these soft tresses, the caddis worm, which observes a vegetarian diet, will find at one and the same time the wherewithal to build and eat. On the other hand, we have a little faggot of bits of wood, very dry, equal in length and each possessing the thickness of a good sized pin. The two sorts of building material lie side by side, mingling their threads and sticks. The animal can make its choice from the lump.

A few hours later, having recovered from the shock of losing its sheath, the caddis worm sets to work to manufacture a new one. It settles across a bunch of tangled rootlets, which are brought together by the builder's legs and more or less arranged by the undulating movement of the hinder part. This gives a kind of incoherent and ill defined suspended belt, a narrow hammock with a number of loose catches; for the various bits of which it is made up are respected by the teeth and extended from place to place beyond the main cords of the roots. Here, without much trouble, is the support, suitably fixed by natural moorings. A few threads of silk, casually distributed, make the frail combination a trifle more secure.

And now to the work of building. Supported by the suspended belt, the caddis worm stretches itself and thrusts out its middle legs, which, being longer than the others, are the grapnels intended to seize things at a distance. It meets a bit of root, fastens on to it, climbs above the point gripped, as though it were measuring the piece to a requisite length, and then, with the fine scissors of its mandibles, cuts the string.

There is at once a brief recoil, which brings the animal back to the level of the hammock. The bit detached lies across the worm's chest, held in its forelegs, which turn it, twist it, wave it about, lay it down, lift it up, as though trying for the best position. Those forelegs make admirably dexterous arms. Being less long than the other two pairs, they are brought into immediate contact with those primordial implements, the mandibles and the spinneret. Their delicate terminal jointing, with a movable and crooked finger, is the caddis worm's equivalent of our hand. They are the working legs. The second pair, which are exceptionally long, serve to spear distant materials and to give the worker a firm footing when measuring a piece and cutting it with the pliers. Lastly, the hind legs, of medium length, afford a support when the others are busy.

The caddis worm, I was saying, with the piece which it has removed held crosswise to its chest, retreats a little way along its suspended hammock until the spinneret is level with the support furnished by the close tangle of rootlets. With a quick movement, it shifts its burden, gets it as nearly by the middle as it can, so that the two ends stick out equally on either side, and chooses the spot to place it, whereupon the spinneret sets to work at once, while the little fore legs hold the scrap of root motionless in its transversal position. The soldering is effected with a touch of silk in the middle of the bit and along a certain distance to the right and left, as far as the bending of the head permits.

Without delay, other sticks are speared in like manner at a distance, cut off and placed in position. As the immediate neighborhood is stripped, the material is gathered at a yet greater distance and the caddis worm bends even farther from its support, which now holds only its last few segments. It is a curious gymnastic display, that of this soft, hanging spine turning and swaying, while the grapnels feel in every direction for a thread.

All this labor results in a sort of casing of little white cords. The work lacks firmness and regularity. Nevertheless, judging by the builder's methods, I can see that the building would not be devoid of merit if the materials gave it a better chance. The caddis worm estimates the size of its pieces very fairly; it cuts them all to nearly the same length; it always arranges them crosswise on the margin of the case; it fixes them by the middle.

Nor is this all: the manner of working helps the general arrangement considerably. When the bricklayer is building the narrow shaft of a factory chimney, he stands in the center of his turret and turns round and round while gradually laying new rows. The caddis worm acts in the same way. It twists round in its sheath; it adopts without inconvenience whatever position it pleases, so as to bring its spinneret full face with the point to be gummed. There is no straining of the neck to left or right, no throwing back of the head to reach points behind. The animal has constantly before it, within the exact range of its implements, the place at which the bit is to be fixed. When the piece is soldered, the worm turns a little aside, to a length equal to that of the last soldering, and here, along an extent which hardly ever varies, an extent determined by the swing which its head is able to give, it fixes the next piece.

These several conditions ought to result in a geometrically ordered dwelling, having a regular polygon as an opening. Then how comes it that the cylinder of bits of root is so confused, so clumsily fashioned? The reason is this: the worker possesses talent, but the materials do not lend themselves to accurate work. The rootlets supply stumps of very uneven shape and thickness. They include big and small ones, straight and bent, simple and ramified. To combine all these dissimilar pieces into an orderly whole is hardly possible, all the more so as the caddis worm does not appear to attach very much importance to its cylinder, which is a temporary work, hurriedly constructed to afford a speedy shelter. Matters are urgent;

and very soft fibers, clipped with a bite of the mandibles, are more quickly gathered and more easily put together than joists, which require the patient work of the saw. The inaccurate cylinder, in short, held in position by numerous guy ropes, is a base upon which a solid and definite structure will rise before long. Soon, the original work will crumble to ruins and disappear, whereas the new one, a permanent structure, will even outlast the owner.

The insects reared in a tumbler show yet another method of building the first dwelling. This time, the caddis worm is given a few very leafy stalks of pond weed (Potamogeton densum) and a bundle of small dry twigs. It perches on a leaf, which the nippers of the mandibles cut half across. The portion left untouched will act as a lanyard and give the necessary steadiness to the early operations.

From an adjoining leaf a section is cut out entirely, an angular and good sized piece. There is plenty of material and no need for economy. The piece is soldered with silk to the strip which was not wholly cut off. The result of three or four similar operations is to surround the Caddis worm with a conical bag, whose wide mouth is scalloped with pointed and very irregular notches. The work of the nippers continues; fresh pieces are fixed, from one to another, inside the funnel, not far from the edge, so that the bag lengthens, tapers and ends by wrapping the animal in a light and floating drapery.

Thus clad for the time being, either in the fine silk of the pond weed or in the linsey-woolsey supplied by the roots of the watercress, the caddis worm begins to think of building a more solid sheath. The present casing will serve as a foundation for the stronger building. But the necessary materials are seldom near at hand: you have to go and fetch them, you have to move your position, an effort which has been avoided until now. With this object, the caddis worm cuts its moorings, that is to say, the rootlets which keep the cylinder fixed, or else the half-severed leaf of pond weed on which the cone-shaped bag has come into being.

The worm is now free. The smallness of the artificial pond, the tumbler, soon brings it into touch with what it is seeking. This is a little faggot of dry twigs, which I have selected of equal length and of slight thickness. Displaying greater care than it did when treating the slender roots, the carpenter measures out the requisite length on the joist. The distance to which it has to extend its body in order to reach the point where the break will be made tells it pretty accurately what length of stick it wants.

The piece is patiently sawn off with the mandibles; it is next taken in the fore legs and held crosswise below the neck. The backward movement which brings the caddis worm home also brings the bit of twig to the edge of the tube. Thereupon, the methods employed in working with the scraps of root are renewed in precisely the same manner. The sticks are scaffolded to the regulation height, all alike in length, amply soldered in the middle and free at either end.

With the picked materials provided, the carpenter has turned out a work of some elegance. The joists are all arranged crosswise, because this way is the handiest for carrying the sticks and putting them in position; they are fixed by the middle, because the two arms that hold the stick while the spinneret does its work require an equal grasp on either side; each soldering covers a length which is seen to be practically invariable, because it is equal to the width described by the head in bending first to this side and then to that when the silk is emitted; the whole assumes a polygonal shape, not far removed from a rectilinear pentagon, because, between laying one piece and the next, the caddis worm turns by the width of an arc corresponding with the length of a soldering. The regularity of the method produces the regularity of the work; but it is essential, of course, that the materials should lend themselves to precise coordination.

In its natural pond, the caddis worm does not often have at its disposal the picked joists which I give it in the tumbler. It comes across something of everything; and that something of everything it employs as it finds it. Bits of wood, large seeds, empty shells, stubble stalks, shapeless fragments are used in the building for better or for worse, just as they occur, without being trimmed by the saw; and this jumble, the result of chance, results in a shockingly faulty structure.

The caddis worm does not forget its talents; but it lacks choice pieces. Give it a proper timber yard and it at once reverts to correct architecture, of which it carries the plans within itself. With small, dead pond snails, all of the same size, it fashions a splendid patchwork scabbard; with a cluster of slender roots, reduced by rotting to their stiff, straight, woody axis, it manufactures pretty specimens of wicker work which could serve as models to our basket makers.

Let us watch it at work when it is unable to use its favorite joist. There is no point in giving it clumsy building stones; that would only bring us back to the uncouth sheaths. Its propensity to make use of soaked seeds, those of the iris, for instance, suggests that I might try grains. I select rice, which, because of its hardness, will be tantamount to wood and, because of its clean whiteness and its oval shape, will lend itself to artistic masonry.

Obviously, my denuded caddis worms cannot start their work with bricks of this kind. Where would they fix their first layer? They must have a foundation, quick and easy to build. This is once more supplied by a temporary cylinder of watercress roots. On this support follow the grains of rice, which, grouped one atop the other, straight or slanting, end by giving a magnificent turret of ivory. Next to the sheaths made of tiny snail shells, this is the prettiest thing with which the caddis worm's industry has furnished me. A fine sense of order has returned, because the materials, regular and of identical character, have cooperated with the correct method of the worker.

The two demonstrations are enough. Sticks and grains of rice make it plain that the caddis worm is not the bungler that one would expect from the monstrous buildings in the pond. Those Cyclopean piles, those mad conglomerations, are the inevitable results of chance finds, which are used for the best because there is no choice. The water carpenter has an art of its own, has method and rules of symmetry. When well served by fortune, it is quite able to turn out good work; when ill-served, it acts like others: the work which it turns out is bad. Poverty makes for ugliness.

There is another matter wherein the caddis worm deserves our attention. With a perseverance which repeated trials do not tire, it makes itself a new tube when I strip it. This is opposed to the habits of the generality of insects, which do not recommence the thing once done, but simply continue it according to the usual rules, taking no account of the ruined or vanished portions. The caddis worm is a striking exception: it starts again. Whence does it derive this capacity?

I begin by learning that, given a sudden alarm, it readily leaves its scabbard. When I go fishing for caddis worms, I put them in tin boxes, containing no other moisture than that wherewith my catches are soaked. I heap them up loosely, to avoid any grievous tumult and to fill the space at my disposal as best I may. I take no further precaution. This is enough to keep the caddis worms in good condition during the two or three hours which I devote to fishing and to walking home.

On my return, I find that a number of them have left their houses. They are swarming naked among the empty scabbards and those still occupied by their inhabitants. It is a pitiful sight to see these evicted ones dragging their bare abdomens and their frail respiratory threads over the bristling sticks. There is no great harm done, however; and I empty the whole lot into the glass pond.

Not one resumes possession of an unoccupied sheath. Perhaps it would take them too long to find one of the exact size. They think it better to abandon the old clouts and to manufacture cases new from top to bottom. The process is a rapid one. By the next day, with the materials wherein the glass trough abounds—bundles of twigs and tufts of watercress—all the denuded worms have made themselves at least a temporary home in the form of a tube of rootlets.

The lack of water, combined with the excitement of the crowding in the boxes, has upset my captives greatly; and, scenting a grave peril, they have made off hurriedly, doffing the cumbersome jacket, which is difficult to carry. They have stripped themselves so as to flee with greater ease. The alarm cannot have been due to me: there are not many simpletons like myself who are interested in the affairs of the pond; and the caddis worm has not been cautioned against their tricks. The sudden desertion of the crib has certainly some other reason than man's molestations.

I catch a glimpse of this reason, the real one. The glass pond was originally occupied by a dozen Dytisci, or water beetles, whose diving performances are so curious to watch. One day, meaning no harm and for want of a better receptacle, I fling among them a couple of handfuls of caddis worms. Blunderer that I am, what have I done! The corsairs, hiding in the rugged corners of the rock work, at once perceive the windfall. They rise to the surface with great strokes of their oars; they hasten and fling themselves upon the crowd of carpenters. Each pirate grabs a sheath by the middle and strives to rip it open by tearing off shells and sticks. While this ferocious enucleation continues with the object of reaching the dainty morsel contained within, the caddis worm, close pressed, appears at the mouth of the sheath, slips out and quickly decamps under the eyes of the Dytiscus, who appears to notice nothing.

I have said before that the trade of killing can dispense with intelligence. The brutal ripper of sheaths does not see the little white sausage that slips between his legs, passes under his fangs and madly flees. He continues to tear away the outer case and to tug at the silken lining. When the breach is made, he is quite crestfallen at not finding what he expected.

Poor fool! Your victim went out under your nose and you never saw it. The worm has sunk to the bottom and taken refuge in the mysteries of the rock work. If things were happening in the large expanse of a pond, it is clear that, with their system of expeditious removals, most of the lodgers would escape scot-free. Fleeing to a distance and recovering from the sharp alarm, they would build themselves a new scabbard and all would be over until the next attack, which would be baffled afresh by the selfsame trick.

In my narrow trough, things take a more tragic turn. When the sheaths are done for, when the caddis worms that are too slow in making off have been eaten up, the Water beetles return to the rockery at the bottom. Here, sooner or later, there are lamentable happenings. The naked fugitives are discovered and, succulent morsels that they are, are forthwith torn to pieces and devoured. Within twenty-four hours, not one of my band of caddis worms is left alive. In order to continue my studies, I had to lodge the water beetles elsewhere.

Under natural conditions, the caddis worm has its persecutors, the most formidable of whom appears to be the Water beetle. When we consider that, to thwart the brigand's attacks, it has invented the idea of quitting its scabbard with all speed, its tactics are certainly most appropriate; but, in that case, an exceptional condition becomes obligatory, namely, the capacity for recommencing the work. This most unusual gift of recommencing it possesses in a high measure. I am ready to see its origin in the persecutions of the Dytiscus and other pirates. Necessity is the mother of industry.

Certain caddis worms, of the Sericostoma and Leptocerus species, clothe themselves in grains of sand and do not leave the bed of the stream. On a clear bottom, swept by the current, they walk about from one bank of verdure to the other and do not think of coming to the surface to float and sail in the sunlight. The collectors of sticks and shells are more highly privileged. They can remain on the level of the water indefinitely, with no other support than their skiff, can rest in unsubmersible flotillas and can even shift their place by working the rudder.

To what do they owe this privilege? Are we to look upon the bundle of sticks as a sort of raft whose density is less than that of the water? Can the shells, which are always empty and able to contain a few bubbles of air in their spiral, he floats? Can the big joists, which break in so ugly a fashion the none too great regularity of the work, serve to buoy up the over-heavy raft? In short, is the caddis worm versed in the laws of equilibrium and does it choose its pieces, now lighter and now heavier as the case may be, so as to constitute a whole that is capable of floating? The following facts are a refutation of any such hydrostatic calculations in the animal.

I remove a number of caddis worms from their sheaths and submit these, as they are, to the test of water. Whether formed wholly of fibrous remnants or of mixed materials, not one of them floats. The scabbards made of shells go to the bottom with the swiftness of a bit of gravel; the others sink gently. I experiment with the separate materials one by one. No shell remains on the surface, not even among the Planorbes, which a many-whorled spiral ought, one would think, to keep afloat. The fibrous remnants must be divided into two categories. The first, darkened by time and soaked with moisture, sink to the bottom. These are the most plentiful. The second, considerably fewer in number, of more recent date and less saturated with water, float very well. The general result is immersion, as in the case of the intact scabbards. I may add that the animal, when removed from its tube, is also unable to float.

Then how does the caddis worm manage to remain on the surface without the support of the grasses, considering that itself and its sheath are both heavier than water? Its secret is soon revealed. I place a few high and dry on a sheet of blotting paper, which will absorb the excess of liquid unfavorable to successful observation. Outside its natural environment, the animal moves about violently and restlessly. With its body half out of the scabbard, this time composed entirely of fibrous matter, it clutches with its feet at the supporting plane. Then, contracting itself, it draws the scabbard towards it, half-raising it and sometimes even making it assume a vertical position. Even so do the Bulimi move along, lifting their shell as they complete each crawling step.

After a couple of minutes in the free air, I replace the caddis worm in the water. This time, it floats, but like a cylinder with too much weight below. The sheath remains vertical, with its hinder orifice level with the water. Soon, an air bubble escapes from the orifice. Deprived of this buoy, the skiff at once goes down.

The result is the same with the caddis worms in shell casings. At first, they float, straight up on end, and then dip under and sink, faster than the others, after sending out an air bubble or two through the back window.

That is enough: the secret is out. When cased in wood or in shells, the caddis worms, which are always heavier than water, are able to keep on the surface by means of a temporary air balloon which decreases the density of the whole structure.

This apparatus works in the simplest manner. Consider the rear of the sheath. It is truncated, wide open and supplied with a membranous partition, the work of the spinneret. A round hole occupies the center of this screen. Beyond it lies the interior of the scabbard, which is smoothly lined and wadded with satin, however rough the exterior may be. Armed at the stern with two hooks which bite into the silky lining, the animal is able to move backwards and forwards at will inside the cylinder, to fix its grapnels at whatever point it pleases and thus to keep a hold on the cylinder while the six legs and the fore part are outside.

When at rest, the body remains indoors entirely and the grub occupies the whole of the tube. But let it contract ever so little towards the front, or, better still, let it stick out a part of its body: a vacuum is formed behind this sort of piston, which may be compared with that of a pump. Thanks to the rear window, a valve without a plug, this vacuum at once fills, thus renewing the aerated water around the gills, a soft fleece of hairs distributed over the back and belly.

The piston stroke affects only the work of breathing; it does not alter the density, makes hardly any change in that which is heavier than water. To lighten the weight, the caddis worm must first rise to the surface. With this object, it scales the grasses of one support after the other; it clambers up, sticking to its purpose in spite of the drawback of its faggot dragging through the tangle. When it has reached the goal, it lifts the rear end a little above the water and gives a stroke of the piston. The vacuum thus obtained fills with air. That is enough: skiff and boatman are in a position to float. The now useless support of the grasses is abandoned. The time has come for evolutions on the surface, in the glad sunlight.

The caddis worm possesses no great talent as a navigator. To turn round, to tack about, to shift its place slightly by a backward movement is all that it can do; and even that it does very clumsily. The front part of the body, sticking out of the case, acts as a rudder. Three or four times over, it rises abruptly, bends, comes down again and strikes the water. These paddle strokes, repeated at intervals, carry the unskilled oarsman to fresh latitudes. It becomes a voyage on the right seas when the crossing measures a hand's breadth.

However, tacking on the surface of the water affords the caddis worm no pleasure. It prefers to twitter in one spot, to remain stationary in flotillas. When the time comes to return to the quiet of the mud bed at the bottom, the animal, having had enough of the sun, draws itself wholly into its sheath again and, with a piston stroke, expels the air from the back room. The normal density is restored and it sinks slowly to the bottom.

We see, therefore, that the caddis worm has not to trouble about hydrostatics when building its scabbard. In spite of the incongruity of its work, in which the bulky and less dense portions seem to balance the more solid, concentrated part, it is not called upon to contrive an equipoise between the light and the heavy. It has other artifices whereby to rise to the surface, to float and to dive down again. The ascent is made by the ladder of the water weeds. The average density of the sheath is of no importance, so long as the burden to be dragged is not beyond the animal's strength. Besides, the weight of the load is greatly reduced when moved in the water.

The admission of a bubble of air into the back chamber, which the animal ceases to occupy, allow it, without further to-do, to remain for an indefinite period on the surface. To dive down again, the caddis worm has only to retreat entirely into its sheath. The air is driven out; and the canoe, resuming its mean density, a greater specific density than that of water, goes under at once and descends of its own accord.

There is, therefore, no choice of materials on the builder's part, no nice calculation of equilibrium, save for one condition, that no stony matter be admitted. That apart, everything serves, large and small, joist and shell, seed and billet. Built up at haphazard, all these things make an impregnable wall. One point alone is essential: the weight of the whole must slightly exceed that of the water displaced; if not, there could be no steadiness at the bottom of the pond, without a perpetual anchorage struggling against the pull of the water. In the same manner, quick submersion would be impossible at times when the surface became dangerous and the frightened creature wanted to leave it.

Nor does this important heavier-than-water question call for lucid discernment, seeing that almost the whole of the sheath is constructed at the bottom of the pond, whither all the materials picked up at random, having descended once before, are likely to descend again. In the sheaths, the parts capable of floating are very rare. Without taking their specific levity into account, simply so as not to remain idle, the caddis worm fixed them to its bundle when sporting on the surface of the water.

We have our submarines, in which hydraulic ingenuity displays its highest resources. The caddis worms have theirs, which emerge, float on the surface, dip down and even stop at mid-depth by releasing gradually their surplus air. And this apparatus, so perfectly balanced, so skilful, requires no knowledge on the part of its constructor. It comes into being of itself, in accordance with the plans of the universal harmony of things.

CHAPTER IX. THE GREENBOTTLES

I have wished for a few things in my life, none of them capable of interfering with the common weal. I have longed to possess a pond, screened from the indiscretion of the passers by, close to my house, with clumps of rushes and patches of duckweed. Here, in my leisure hours, in the shade of a willow, I should have meditated upon aquatic life, a primitive life, easier than our own, simpler in its affections and its brutalities. I should have watched the unalloyed happiness of the mollusk, the frolics of the Whirligig, the figure-skating of the Hydrometra [a water bug known as the Pond skater], the dives of the Dytiscus beetle, the veering and tacking of the Notonecta [the water boatman], who, lying on her back, rows with two long oars, while her short forelegs, folded against her chest, wait to grab the coming prey. I should have studied the eggs of the Planorbis, a glairy nebula wherein focuses of life are condensed even as suns are condensed in the nebulae of the heavens. I should have admired the nascent creature that turns, slowly turns in the orb of its egg and describes a volute, the draft, perhaps, of the future shell. No planet circles round its center of attraction with greater geometrical accuracy.

I should have brought back a few ideas from my frequent visits to the pond. Fate decided otherwise: I was not to have my sheet of water. I have tried the artificial pond, between four panes of glass. A poor shift! Our laboratory aquariums are not even equal to the print left in the mud by a mule's hoof, when once a shower has filled the humble basin and life has stocked it with its marvels.

In spring, with the hawthorn in flower and the crickets at their concerts, a second wish often came to me. Along the road, I light upon a dead mole, a snake killed with a stone, victims both of human folly. The mole was draining the soil and purging it of its vermin. Finding him under his spade, the laborer broke his back for him and flung him over the hedge. The snake, roused from her slumber by the soft warmth of April, was coming into the sun to shed her skin and take on a new one. Man catches sight of her: 'Ah, would you?' says he. 'See me do something for which the world will thank me!'

And the harmless beast, our auxiliary in the terrible battle which husbandry wages against the insect, has its head smashed in and dies.

The two corpses, already decomposing, have begun to smell. Whoever approaches with eyes that do not see turns away his head and passes on. The observer stops and lifts the remains with his foot; he looks. A world is swarming underneath; life is eagerly consuming the dead. Let us replace matters as they were and leave death's artisans to their task. They are engaged in a most deserving work.

To know the habits of those creatures charged with the disappearance of corpses, to see them busy at their work of disintegration, to follow in detail the process of transmutation that makes the ruins of what has lived return apace into life's treasure house: these are things that long haunted my mind. I regretfully left the mole lying in the dust of the road. I had to go, after a glance at the corpse and its harvesters. It was not the place for philosophizing over a stench. What would people say who passed and saw me!

And what will the reader himself say, if I invite him to that sight? Surely, to busy one's self with those squalid sextons means soiling one's eyes and mind? Not so, if you please! Within the domain of our restless curiosity, two questions stand out above all others: the question of the beginning and the question of the end. How does matter unite in order to assume life? How does it separate when returning to inertia? The pond, with its Planorbis eggs turning round and round, would have given us a few data for the first problem; the Mole, going bad under conditions not too repulsive, will tell us something about the second: he will show us the working of the crucible wherein all things are melted to begin anew. A truce to nice delicacy! Odi profanum vulgus et arceo; hence, ye profane: you would not understand the mighty lesson of the rag tank.

I am now in a position to realize my second wish. I have space, air and quiet in the solitude of the harmas. None will come here to trouble me, to smile or to be shocked at my investigations. So far, so good; but observe the irony of things: now that I am rid of passers by, I have to fear my cats, those assiduous prowlers, who, finding my preparations, will not fail to spoil and scatter them. In anticipation of their misdeeds, I establish workshops in midair, whither none but genuine corruption agents can come, flying on their wings. At different points in the enclosure, I plant reeds, three by three, which, tied at their free ends, form a stable tripod. From each of these supports, I hang, at a man's height, an earthenware pan filled with fine sand and pierced at the bottom with a hole to allow the water to escape, if it should rain. I garnish my apparatus with dead bodies. The snake, the lizard, the toad receive the preference, because of their bare skins, which enable me better to follow the first attack and the work of the invaders. I ring the changes with furred and feathered beasts. A few children of the neighborhood, allured by pennies, are my regular purveyors. Throughout the good season, they come running triumphantly to my door, with a snake at the end of a stick, or a lizard in a cabbage leaf. They bring me the rat caught in a trap, the chicken dead of the pip, the mole slain by the gardener, the kitten killed by accident, the rabbit poisoned by some weed. The business proceeds to the mutual satisfaction of sellers and buyer. No such trade had ever been known before in the village nor ever will be again.

April ends; and the pans rapidly fill. An ant, ever so small, is the first arrival. I thought I should keep this intruder off by hanging my apparatus high above the ground: she laughs at my precautions. A few hours after the deposit of the morsel, fresh still and possessing no appreciable smell, up comes the eager picker-up of trifles, scales the stems of the tripod in processions and starts the work of dissection. If the joint suits her, she even goes to live in the sand of the pan and digs herself temporary platforms in order to work the rich find more at her ease.

All through the season, from start to finish, she will always be the promptest, always the first to discover the dead animal, always the last to beat a retreat when nothing more remains than a heap of little bones bleached by the sun. How does the vagabond, passing at a distance, know that, up there, invisible, high on the gibbet, there is something worth going for? The others, the real knackers, wait for the meat to go bad; they are informed by the strength of the effluvia. The ant, gifted with greater powers of scent, hurries up before there is any stench at all. But, when the meat, now two days old and ripened by the sun, exhales its flavor, soon the master ghouls appear upon the scene: Dermestes [bacon beetles, small flesh-eating beetles] and Saprini [exceedingly small flesh-eating beetles], Silphae [carrion beetles] and Necrophori [burying beetles], flies and Staphylini [rove beetles], who attack the corpse, consume it and reduce it almost to nothing. With the ant alone, who each time carries off a mere atom, the sanitary operation would take too long; with them, it is a quick business, especially as certain of them understand the process of chemical solvents.

These last, who are high class scavengers, are entitled to first mention. They are flies, of many various species. If time permitted, each of those strenuous ones would deserve a special examination; but that would weary the patience of both the reader and the observer. The habits of one will give us a summary notion of the habits of the rest. We will therefore confine ourselves to the two principal subjects, namely, the Luciliae, or greenbottles, and the Sarcophagae, or grey flesh flies.

The Luciliae—flies that glitter—are magnificent flies known to all of us. Their metallic luster, generally a golden green, rivals that of our finest beetles, the Rosechafers, Buprestes and leaf beetles. It gives one a shock of surprise to see so rich a garb adorn those workers in putrefaction. Three species frequent my pans: Lucilia Caesar, LIN., L. cadaverina, LIN., and L.

cuprea, ROB. The first two, both of whom are gold-green, are plentiful; the third, who sports a coppery luster, is rare. All three have red eyes, set in a silver border.

Lucilia Caesar is larger than L. cadaverina and also more forward in her business. I catch her in labor on the 23rd of April. She has settled in the spinal canal of a neck of mutton and is laying her eggs on the marrow. For more than an hour, motionless in the gloomy cavity, she goes on packing her eggs. I can just see her red eyes and her silvery face. At last, she comes out. I gather the fruit of her labor, an easy matter, for it all lies on the marrow, which I extract without touching the eggs.

A census would seem important. To take it at once is impracticable: the germs form a compact mass, which would be difficult to count. The best thing is to rear the family in a jar and to reckon by the pupae buried in the sand. I find a hundred and fifty-seven. This is evidently but a minimum; for Lucilia Caesar and the others, as the observations that follow will tell me, lay in packets at repeated intervals. It is a magnificent family, promising a fabulous legion to come.

The greenbottles, I was saying, break up their laying into sections. The following scene affords a proof of this. A Mole, shrunk by a few days' evaporation, lies spread upon the sand of the pan. At one point, the edge of the belly is raised and forms a deep arch. Remark that the Greenbottles, like the rest of the flesh eating flies, do not trust their eggs to uncovered surfaces, where the heat of the sun's rays might endanger the existence of the delicate germs. They want dark hiding places. The favorite spot is the lower side of the dead animal, when this is accessible.

In the present case, the only place of access is the fold formed by the edge of the belly. It is here and here alone that this day's mothers are laying. There are eight of them. After exploring the piece and recognizing its good quality, they disappear under the arch, first this one, then that, or else several at a time. They remain under the Mole for a considerable while. Those outside wait, but go repeatedly to the threshold of the cavern to take a look at what is happening within and see whether the earlier ones have finished. These come out at last, perch on the animal and wait in their turn. Others at once take their place in the recesses of the cave. They remain there for some time and then, having done their business, make room for more mothers and come forth into the sunlight. This going in and out continues throughout the morning.

We thus learn that the laying is effected by periodical emissions, broken with intervals of rest. As long as she does not feel ripe eggs coming to her oviduct, the greenbottle remains in the sun, hovering to and fro and sipping modest mouthfuls from the carcass. But, as soon as a fresh stream descends from her ovaries, quick as lightning she makes for a propitious site whereon to deposit her burden. It appears to be the work of several days thus to divide the total laying and to distribute it at different points.

I carefully raise the animal under which these things are happening. The egg laying mothers do not disturb themselves; they are far too busy. Their ovipositor extended telescope fashion, they heap egg upon egg. With the point of their hesitating, groping instrument, they try to lodge each germ, as it comes, farther into the mass. Around the serious, red-eyed matrons, the Ants circle, intent on pillage. Many of them make off with a greenbottle egg between their teeth. I see some who, greatly daring, effect their theft under the ovipositor itself. The layers do not put themselves out, let the ants have their way, remain impassive. They know their womb to be rich enough to make good any such larceny.

Indeed, what escapes the depredations of the ants promises a plenteous brood. Let us come back a few days later and lift the mole again. Underneath, in a pool of sanies, is a surging mass of swarming sterns and pointed heads, which emerge, wriggle and dive in again. It suggests a seething billow. It turns one's stomach. It is horrible, most horrible. Let us steel ourselves against the sight: it will be worse elsewhere.

Here is a fat snake. Rolled into a compact whorl, she fills the whole pan. The greenbottles are plentiful. New ones arrive at every moment and, without quarrel or strife, take their place among the others, busily laying. The spiral furrow left by the reptile's curves is the favorite spot. Here alone, in the narrow space between the folds, are shelters against the heat of the sun. The glistening Flies take their places, side by side, in rows; they strive to push their abdomen and their ovipositor as far forward as possible, at the risk of rumpling their wings and cocking them towards their heads. The care of the person is neglected amid this serious business. Placidly, with their red eyes turned outwards, they form a continuous cordon. Here and there, at intervals, the rank is broken; layers leave their posts, come and walk about upon the snake, what time their ovaries ripen for another emission, and then hurry back, slip into the rank and resume the flow of germs. Despite these interruptions, the work of breeding goes fast. In the course of one morning, the depths of the spiral furrow are hung with a continuous white bark, the heaped up eggs. They come off in great slabs, free of any stain; they can be shoveled up, as it were, with a paper scoop. It is a propitious moment if we wish to follow the evolution at close quarters. I therefore gather a profusion of this white manna and lodge it in glass tubes, test tubes and jars, with the necessary provisions.

The eggs, about a millimeter long, are smooth cylinders, rounded at both ends. They hatch within twenty-four hours. The first question that presents itself is this: how do the greenbottle grubs feed? I know quite well what to give them, but I do not in the least see how they manage to consume it. Do they eat, in the strict sense of the word? I have reasons to doubt it.

Let us consider the grub grown to a sufficient size. It is the usual fly larva, the common maggot, shaped like an elongated cone, pointed in front, truncated behind, where two little red spots show, level with the skin: these are the breathing holes. The front, which is called the head by stretching a word—for it is little more than the entrance to an intestine—the front is armed with two little black hooks, which slide in a translucent sheath, project a little way outside and go in turn by turn. Are we to look upon these as mandibles? Not at all, for, instead of having their points facing each other, as would be required in a real mandibular apparatus, the two hooks work in parallel directions and never meet. What they are is ambulatory organs, grapnels assisting locomotion, which give a purchase on the plane and enable the animal to advance by means of repeated contractions. The maggot walks with the aid of what a superficial examination would pronounce to be a machine for eating. It carries in its gullet the equivalent of the climber's alpenstock.

Let us hold it, on a piece of flesh, under the lens. We shall see it walking about, raising and lowering its head and, each time, stabbing the meat with its pair of hooks. When stationary, with its crupper at rest, it explores space with a continual bending of its fore part; its pointed head pokes about, jabs forward, goes back again, producing and withdrawing its black mechanism. There is a perpetual piston play. Well, look as carefully and conscientiously as I please, I do not once see the weapons of the mouth tackle a particle of flesh that is torn away and swallowed. The hooks come down upon the meat at every moment, but never take a visible mouthful from it. Nevertheless, the grub waxes big and fat. How does this singular consumer, who feeds without eating, set about it? If he does not eat, he must drink; his diet is soup. As meat is a compact substance, which does not liquefy of its own accord, there must, in that case, be a certain recipe to dissolve it into a fluid broth. Let us try to surprise the maggot's secret.

In a glass tube, sealed at one end, I insert a piece of lean flesh, the size of a walnut, which I have drained of its juices by squeezing it in blotting paper. On the top of this, I place a few slabs of greenbottle eggs collected a moment ago from the snake in my earthen pan. The number of germs is, roughly, two hundred. I close the tube with a cotton plug, stand it upright, in a shady corner of my study, and leave things to take their course. A control tube, prepared like the first, but not stocked with maggots, is placed beside it.

As early as two or three days after the hatching, I obtain a striking result. The meat, which was thoroughly drained by the blotting paper, has become so moist that the young vermin leave a wet mark behind them as they crawl over the glass. The swarming brood creates a sort of mist with the crossing and criss-crossing of its trails. The control tube, on the contrary, keeps dry, proving that the moisture in which the worms move is not due to a mere exudation from the meat.

Besides, the work of the maggot becomes more and more evident. Gradually, the flesh flows in every direction like an icicle placed before the fire. Soon, the liquefaction is complete. What we see is no longer meat, but fluid Liebig's extract. If I overturned the tube, not a drop of it would remain.

Let us clear our minds of any idea of solution by putrefaction, for in the second tube a piece of meat of the same kind and size has remained, save for color and smell, what it was at the start. It was a lump and it is a lump, whereas the piece treated by the worms runs like melted butter. Here we have maggot chemistry able to rouse the envy of physiologists when studying the action of the gastric juice.

I obtain better results still with hard-boiled white of egg. When cut into pieces the size of a hazel nut and handed over to the greenbottle's grubs, the coagulated albumen dissolves into a colorless liquid which the eye might mistake for water. The fluidity becomes so great that, for lack of a support, the worms perish by drowning in the broth; they are suffocated by the immersion of their hind part, with its open breathing holes. On a denser liquid, they would have kept at the surface; on this, they cannot.

A control tube, filled in the same way, but not colonized, stands beside that in which the strange liquefaction takes place. The hardboiled white of egg retains its original appearance and consistency. In course of time, it dries up, if it does not turn moldy; and that is all.

The other quaternary compounds performing the same functions as albumen—the gluten of cereals, the fibrin of blood, the casein of cheese and the legumin of chickpeas—undergo a similar modification, in varying degrees. Fed, from the moment of leaving the egg, on any one of these substances, the worms thrive very well, provided that they escape drowning when the gruel becomes too clear; they would not fare better on a corpse. And, as a general rule, there is not much danger of going under: the matter only half liquefies; it becomes a running pea soup, rather than an actual fluid.

Even in this imperfect case, it is obvious that the greenbottle grubs begin by liquefying their food. Incapable of taking solid nourishment, they first transform the spoil into running matter; then, dipping their heads into the product, they drink, they slake their thirst, with long sups. Their dissolvent, comparable in its effects with the gastric juice of the higher animals, is, beyond a doubt, emitted through the mouth. The piston of the hooks, continually in movement, never ceases spitting it out in infinitesimal doses. Each spot touched receives a grain of some subtle pepsin, which soon suffices to make that spot run in every direction. As digesting, when all is said, merely means liquefying, it is no paradox to assert that the maggot digests its food before swallowing it.

These experiments with my filthy, evil smelling tubes have given me some delightful moments. The worthy Abbe Spallanzani must have known some such when he saw pieces of raw meat begin to run under the action of the gastric juice which he took, with pellets of sponge, from the stomachs of crows. He discovered the secrets of digestion; he realized in a glass tube the hitherto unknown labors of gastric chemistry. I, his distant disciple, behold once more, under a most unexpected aspect, what struck the Italian scientist so forcibly. Worms take the place of the crows. They slaver upon meat, gluten, albumen; and those substances turn to fluid. What our stomach does within its mysterious recesses the maggot achieves outside, in the open air. It first digests and then imbibes.

When we see it plunging into the carrion broth, we even wonder if it cannot feed itself, at least to some extent, in a more direct fashion. Why should not its skin, which is one of the most delicate, be capable of absorbing? I have seen the egg of the sacred beetle and other dung beetles growing considerably larger—I should like to say, feeding—in the thick atmosphere of the hatching chamber. Nothing tells us that the grub of the greenbottle does not adopt this method of growing. I picture it capable of feeding all over the surface of its body. To the gruel absorbed by the mouth it adds the balance of what is gathered and strained through the skin. This would explain the need for provisions liquefied beforehand.

Let us give one last proof of this preliminary liquefaction. If the carcass—mole, snake or another—left in the open air have a wire gauze cover placed over it, to keep out the flies, the game dries under a hot sun and shrivels up without appreciably wetting the sand on which it lies. Fluids come from it, certainly, for every organized body is a sponge swollen with water; but the liquid discharge is so slow and restricted in quantity that the heat and the dryness of the air disperse it as it appears, while the underlying sand remains dry, or very nearly so. The carcass becomes a sapless mummy, a mere bit of leather. On the other hand, do not use the wire gauze cover, let the flies do their work unimpeded; and things forthwith assume another aspect. In three or four days, an oozing sanies appears under the animal and soaks the sand to some distance.

I shall never forget the striking spectacle with which I conclude this chapter. This time, the dish is a magnificent Aesculapius' snake, a yard and a half long and as thick as a wide bottleneck. Because of its size, which exceeds the dimensions of my pan, I roll the reptile in a double spiral, or in two storeys. When the copious joint is in full process of dissolution, the pan becomes a puddle wherein wallow, in countless numbers, the grubs of the greenbottle and those of Sarcophaga carnaria, the Grey or checkered flesh fly, which are even mightier liquefiers. All the sand in the apparatus is saturated, has turned into mud, as though there had been a shower of rain. Through the hole at the bottom, which is protected by a flat pebble, the gruel trickles drop by drop. It is a still at work, a mortuary still, in which the Snake is being drawn off. Wait a week or two; and the whole will have disappeared, drunk up by the sun: naught but the scales and bones will remain on a sheet of mud.

To conclude: the maggot is a power in this world. To give back to life, with all speed, the remains of that which has lived, it macerates and condenses corpses, distilling them into an essence wherewith the earth, the plant's foster mother, may be nourished and enriched.

CHAPTER X. THE GREY FLESH FLIES

Here the costume changes, not the manner of life. We find the same frequenting of dead bodies, the same capacity for the speedy liquefaction of the fleshy matter. I am speaking of an ash-gray fly, the greenbottle's superior in size, with brown streaks on her back and silver gleams on her abdomen. Note also the blood-red eyes, with the hard look of the knacker in them. The language of science knows her as Sarcophaga, the flesh eater; in the vulgar tongue she is the grey flesh fly, or simply the flesh fly.

Let not these expressions, however accurate, mislead us into believing for a moment that the Sarcophagae are the bold company of master tainters who haunt our dwellings, more particularly in autumn, and plant their vermin in our ill-guarded viands. The author of those offences is Calliphora vomitoria, the bluebottle, who is of a stouter build and arrayed in darkest blue. It is she who buzzes against our windowpanes, who craftily besieges the meat safe and who lies in wait in the darkness for an opportunity to outwit our vigilance. The other, the grey fly, works jointly with the greenbottles, who do not venture inside our houses and who work in the sunlight. Less timid, however, than they, should the outdoor yield be small, she will sometimes come indoors to perpetrate her villainies. When her business is done, she makes off as fast as she can, for she does not feel at home with us.

At this moment, my study, a very modest extension of my open air establishments, has become something of a charnel house. The grey fly pays me a visit. If I lay a piece of butcher's meat on the windowsill, she hastens up, works her will on it and retires. No hiding place escapes her notice among the jars, cups, glasses and receptacles of every kind with which my shelves are crowded.

With a view to certain experiments, I collected a heap of wasp grubs, asphyxiated in their underground nests. Stealthily she arrives, discovers the fat pile and, hailing as treasure trove this provender whereof her race perhaps has never made use before, entrusts to it an installment of her family. I have left at the bottom of a glass the best part of a hard-boiled egg from which I have taken a few bits of white intended for the greenbottle maggots. The grey fly takes possession of the remains, recks not of their novelty and colonizes them. Everything suits her that falls within the category of albuminous matters: everything, down to dead silkworms; everything, down to a mess of kidney-beans and chick-peas.

Nevertheless, her preference is for the corpse: furred beast and feathered beast, reptile and fish, indifferently. Together with the greenbottles, she is sedulous in her attendance on my pans. Daily she visits my snakes, takes note of the condition of each of them, savors them with her proboscis, goes away, comes back, takes her time and at last proceeds to business. Still, it is not here, amid the tumult of callers, that I propose to follow her operations. A lump of butcher's meat laid on the window sill, in front of my writing table, will be less offensive to the eye and will facilitate my observations.

Two flies of the genus Sarcophaga frequent my slaughter yard: Sarcophaga carnaria and Sarcophaga haemorrhoidalis, whose abdomen ends in a red speck. The first species, which is a little larger than the second, is more numerous and does the best part of the work in the open air shambles of the pans. It is this fly also who, at intervals and nearly always alone, hastens to the bait exposed on the windowsill.

She comes up suddenly, timidly. Soon she calms herself and no longer thinks of fleeing when I draw near, for the dish suits her. She is surprisingly quick about her work. Twice over—buzz! Buzz!—the tip of her abdomen touches the meat; and the thing is done: a group of vermin wriggles out, releases itself and disperses so nimbly that I have no time to take my lens and count then accurately. As seen by the naked eye, there were a dozen of them. What has become of them? One would think that they had gone into the flesh, at the very spot where they were laid, so quickly have they disappeared. But that dive into a substance of some consistency is impossible to these newborn weaklings. Where are they? I find them more or less everywhere in the creases of the meat; singly and already groping with their mouths. To collect them in order to number them is not practicable, for I do not want to damage them. Let us be satisfied with the estimate made at a rapid glance: there are a dozen or so, brought into the world in one discharge of almost inappreciable length.

Those live grubs, taking the place of the usual eggs, have long been known. Everybody is aware that the flesh flies bring forth living maggots, instead of laying eggs. They have so much to do and their work is so urgent! To them, the instruments of the transformation of dead matter, a day means a day, a long space of time which it is all important to utilize. The greenbottle's eggs, though these are of very rapid development, take twenty-four hours to yield their grubs. The flesh flies save all this time. From their matrix, laborers flow straightway and set to work the moment they are born. With these ardent pioneers of sanitation, there is no rest attendant upon the hatching, there is not a minute lost.

The gang, it is true, is not a numerous one; but how often can it not be renewed! Read Reaumur's description of the wonderful procreating machinery boasted by the Flesh flies. It is a spiral ribbon, a velvety scroll whose nap is a sort of fleece of maggots set closely together and each cased in a sheath. The patient biographer counted the host: it numbers, he tells us, nearly twenty thousand. You are seized with stupefaction at this anatomical fact.

How does the gray fly find the time to settle a family of such dimensions, especially in small packets, as she has just done on my window sill? What a number of dead dogs, moles and snakes must she not visit before exhausting her womb! Will she find them? Corpses of much size do not abound to that extent in the country. As everything suits her, she will alight on other remains of minor importance. Should the prize be a rich one, she will return to it tomorrow, the day after and later still, over and over again. In the course of the season, by dint of packets of grubs deposited here, there and everywhere, she will perhaps end by housing her entire brood. But then, if all things prosper, what a glut, for there are several families born during the year! We feel it instinctively: there must be a check to these generative enormities. Let us first consider the grub. It is a sturdy maggot, easy to distinguish from the greenbottle's by its larger girth and especially by the way in which its body terminates behind. There is here a sudden breaking off, hollowed into a deep cup. At the bottom of this crater are two breathing holes, two stigmata with amber-red tips. The edge of the cavity is fringed with half a score of pointed, fleshy festoons, which diverge like the spikes of a coronet. The creature can close or open this diadem at will by bringing the denticulations together or by spreading them out wide. This protects the air holes which might otherwise be choked up when the maggot disappears in the sea of broth. Asphyxia would supervene, if the two breathing holes at the back became obstructed. During the immersion, the festooned coronet shuts like a flower closing its petals and the liquid is not admitted to the cavity.

Next follows the emergence. The hind part reappears in the air, but appears alone, just at the level of the fluid. Then the coronet spreads out afresh, the cup gapes and assumes the aspect of a tiny flower, with the white denticulations for petals and the two bright red dots, the stigmata at the bottom, for stamens. When the grubs, pressed one against the other, with their heads downwards in the fetid soup, make an unbroken shoal, the sight of those breathing cups incessantly opening and closing, with a little clack like a valve, almost makes one forget the horrors of the charnel yard. It suggests a carpet of tiny Sea anemones. The maggot has its beauties after all.

It is obvious, if there be any logic in things, that a grub so well-protected against asphyxiation by drowning must frequent liquid surroundings. One does not encircle one's hindquarters with a coronet for the sole satisfaction of displaying it. With its apparatus of spokes, the Grey Fly's grub informs us of the dangerous nature of its functions: when working upon a corpse, it runs the risk of drowning. How is that? Remember the grubs of the greenbottle, fed on hard-boiled white of egg. The dish suits them; only, by the action of their pepsin, it becomes so fluid that they die submerged. Because of their hinder stigmata, which are actually on the skin and devoid of any defensive machinery, they perish when they find no support apart from the liquid.

The flesh fly's maggots, though incomparable liquefiers, know nothing of this peril, even in a puddle of carrion broth. Their bulky hind part serves as a float and keeps the air holes above the surface. When, for further investigation, they must needs go under completely, the anemone at the back shuts and protects the stigmata. The grubs of the gray fly are endowed with a life buoy because they are first class liquefiers, ready to incur the danger of a ducking at any moment.

When high and dry on the sheet of cardboard where I place them to observe them at my ease, they move about actively, with their breathing rose widespread and their stigmata rising and falling as a support. The cardboard is on my table, at three steps from an open window, and lit at this time of day only by the soft light of the sky. Well, the maggots, one and all of them, turn in the opposite direction to the window; they hastily, madly take to flight.

I turn the cardboard round, without touching the runaways. This action makes the creatures face the light again. Forthwith, the troop stops, hesitates, takes a half turn and once more retreats towards the darkness. Before the end of the racecourse is reached, I again turn the cardboard. For the second time, the maggots veer round and retrace their steps. Repeat the experiment as often as I will, each time the squad wheels about in the opposite direction to the window and persists in avoiding the trap of the revolving cardboard.

The track is only a short one: the cardboard measures three hand's breadths in length. Let us give more space. I settle the grubs on the floor of the room; with a hair pencil, I turn them with their heads pointing towards the lighted aperture. The moment they are free, they turn and run from the light. With all the speed whereof their cripple's shuffle allows, they cover the tiled floor of the study and go and knock their heads against the wall, twelve feet off, skirting it afterwards, some to the right and some to the left. They never feel far enough away from that hateful illuminated opening.

What they are escaping from is evidently the light, for, if I make it dark with a screen, the troop does not change its direction when I turn the cardboard. It then progresses quite readily towards the window; but, when I remove the screen, it turns tail at once.

That a grub destined to live in the darkness, under the shelter of a corpse, should avoid the light is only natural; the strange part is its very perception. The maggot is blind. Its pointed fore part, which we hesitate to call a head, bears absolutely no trace of any optical apparatus; and the same with every other part of the body. There is nothing but one bare, smooth, white skin. And this sightless creature, deprived of any special nervous points served by ocular power, is extremely sensitive to the light. Its whole skin is a sort of retina, incapable of seeing, of course, but able, at any rate, to distinguish between light and darkness. Under the direct rays of a searching sun, the grub's distress could be easily explained. We ourselves; with our coarse skin, in comparison with that of the maggot, can distinguish between sunshine and shadow without the help of the eyes. But, in the present case, the problem becomes singularly complicated. The subjects of my experiment receive only the diffused light of the sky, entering my study through an open window; yet this tempered light frightens them out of their senses. They flee the painful apparition; they are bent upon escaping at all costs.

Now what do the fugitives feel? Are they physically hurt by the chemical radiations? Are they exasperated by other radiations, known or unknown? Light still keeps many a secret hidden from us and perhaps our optical science, by studying the maggot, might become the richer by some valuable information. I would gladly have gone farther into the question, had I possessed the necessary apparatus. But I have not, I never have had and of course I never shall have the resources which are so useful to the seeker. These are reserved for the clever people who care more for lucrative posts than for fair truths. Let us continue, however, within the measure which the poverty of my means permits.

When duly fattened, the grubs of the flesh flies go underground to transform themselves into pupae. The burial is intended, obviously, to give the worm the tranquillity necessary for the metamorphosis. Let us add that another object of the descent is to avoid the importunities of the light. The maggot isolates itself to the best of its power and withdraws from the garish day before contracting into a little keg. In ordinary conditions, with a loose soil, it goes hardly lower than a hand's breadth down, for provision has to be made for the difficulties of the return to the surface when the insect, now full grown, is impeded by its delicate fly wings. The grub, therefore, deems itself suitably isolated at a moderate depth. Sideways, the layer that shields it from the light is of indefinite thickness; upwards, it measures about four inches. Behind this screen reigns utter darkness, the buried one's delight. This is capital.

What would happen if, by an artifice, the sideward layer were nowhere thick enough to satisfy the grub? Now, this time, I have the wherewithal to solve the problem, in the shape of a big glass tube, open at both ends, about three feet long and less than an inch wide. I use it to blow the flame of hydrogen in the little chemistry lessons which I give my children.

I close one end with a cork and fill the tube with fine, dry, sifted sand. On the surface of this long column, suspended perpendicularly in a corner of my study, I install some twenty Sarcophaga grubs, feeding them with meat. A similar preparation is repeated in a wider jar, with a mouth as broad as one's hand. When they are big enough, the grubs in either apparatus will go down to the depth that suits them. There is no more to be done but to leave them to their own devices.

The worms at last bury themselves and harden into pupae. This is the moment to consult the two apparatus. The jar gives me the answer which I should have obtained in the open fields. Four inches down, or thereabouts, the worms have found a quiet lodging, protected above by the layer through which they have passed and on every side by the thickness of the vessel's contents. Satisfied with the site, they have stopped there.

It is a very different matter in the tube. The least buried of the pupae are half a yard down. Others are lower still; most of them even have reached the bottom of the tube and are touching the cork stopper, an insuperable barrier. These last, we can see, would have gone yet deeper if the apparatus had allowed them. Not one of the score of grubs has settled at the customary halting place; all have traveled farther down the column, until their strength gave way. In their anxious flight, they have dug deeper and ever deeper.

What were they flying from? The light. Above them, the column traversed forms a more than sufficient shelter; but, at the sides, the irksome sensation is still felt through a coat of earth half an inch thick if the descent is made perpendicularly. To escape the disturbing impression, the grub therefore goes deeper and deeper, hoping to obtain lower down the rest which is denied it above. It only ceases to move when worn out with the effort or stopped by an obstacle.

Now, in a soft diffused light, what can be the radiations capable of acting upon this lover of darkness? They are certainly not the simple luminous rays, for a screen of fine, heaped up earth, nearly half an inch in thickness, is perfectly opaque. Then, to alarm the grub, to warn it of the over proximity of the exterior and send it to mad depths in search of isolation, other radiations, known or unknown, must be required, radiations capable of penetrating a screen against which ordinary radiations are powerless. Who knows what vistas the natural philosophy of the maggot might open out to us? For lack of apparatus, I confine myself to suspicions.

To go underground to a yard's depth—and farther if my tube had allowed it—is on the part of the Flesh fly's grub a vagary provoked by unkind experiment: never would it bury itself so low down, if left to its own wisdom. A hand's breadth thickness is quite enough, is even a great deal when, after completing the transformation, it has to climb back to the surface, a laborious operation absolutely resembling the task of an entombed well sinker. It will have to fight against the sand that slips and gradually fills up the small amount of empty space obtained; it will perhaps, without crowbar or pickaxe, have to cut itself a gallery through something tantamount to tufa, that is to say, through earth which a shower has rendered compact. For the descent, the grub has its fangs; for the assent, the fly has nothing. Only that moment come into existence, she is a weakling, with tissues still devoid of any firmness. How does she manage to get out? We shall know by watching a few pupae placed at the bottom of a test-tube filled with earth. The method of the Flesh flies will teach us that of the greenbottles and the other Flies, all of whom make use of the same means.

Enclosed in her pupa, the nascent fly begins by bursting the lid of her casket with a hernia which comes between her two eyes and doubles or trebles the size of her head. This cephalic blister throbs: it swells and subsides by turns, owing to the alternate flux and reflux of the blood. It is like the piston of an hydraulic press opening and forcing back the front part of the keg.

The head makes its appearance. The hydrocephalous monster continues the play of her forehead, while herself remaining stationary. Inside the pupa, a delicate work is being performed: the casting of the white nymphal tunic. All through this operation, the hernia is still projecting. The head is not the head of a fly, but a queer, enormous mitre, spreading at the base into two red skull caps, which are the eyes. To split her cranium in the middle, shunt the two halves to the right and left and send surging through the gap a tumor which staves the barrel with its pressure: this constitutes the Fly's eccentric method.

For what reason does the hernia, once the keg is staved, continue swollen and projecting? I take it to be a waste pocket into which the insect momentarily forces back its reserves of blood in order to diminish the bulk of the body to that extent and to extract it more easily from the nymphal slough and afterwards from the narrow channel of the shell. As long as the operation of the release lasts, it pushes outside all that it is able to inject of its accumulated humors; it makes itself small inside the pupa and swells into a bloated deformity without. Two hours and more are spent in this laborious stripping.

At last, the fly comes into view. The wings, mere scanty stumps, hardly reach the middle of the abdomen. On the outer edge, they have a deep notch similar to the waist of a violin. This diminishes by just so much the surface and the length, an excellent device for decreasing the friction along the earthy column which has next to be scaled. The hydrocephalous one resumes her performance more vigorously than ever; she inflates and deflates her frontal knob. The pounded sand rustles down the insect's sides. The legs play but a secondary part. Stretched behind, motionless, when the piston stroke is delivered, they furnish a support. As the sand descends, they pile it and nimbly push it back, after which they drag along lifelessly until the next avalanche. The head advances each time by a length equal to that of the sand displaced. Each stroke of the frontal swelling means a step forward. In a dry, loose soil, things go pretty fast. A column six inches high is traversed in less than a quarter of an hour.

As soon as it reaches the surface, the insect, covered with dust, proceeds to make its toilet. It thrusts out the blister of its forehead for the last time and brushes it carefully with its front tarsi. It is important that the little pounding engine should be carefully dusted before it is taken inside to form a forehead that will open no more: this lest any grit should lodge in the head. The wings are carefully brushed and polished; they lose their curved notches; they lengthen and spread. Then, motionless on the surface of the sand, the fly matures fully. Let us set her at liberty. She will go and join the others on the Snakes in my pans.

CHAPTER XI. THE BUMBLEBEE FLY

Underneath the wasp's brown paper manor house, the ground is channeled into a sort of drain for the refuse of the nest. Here are shot the dead or weakly larvae which a continual inspection roots out from the cells to make room for fresh occupants; here, at the time of the autumn massacre, are flung the backward grubs; here, lastly, lies a good part of the crowd killed by the first touch of winter. During the rack and ruin of November and December, this sewer becomes crammed with animal matter.

Such riches will not remain unemployed. The world's great law which says that nothing edible shall be wasted provides for the consumption of a mere ball of hair disgorged by the owl. How shall it be with the vast stores of a ruined wasps' nest! If they have not come yet, the consumers whose task it is to salve this abundant wreckage for nature's markets, they will not tarry in coming and waiting for the manna that will soon descend from above. That public granary, lavishly stocked by death, will become a busy factory of fresh life. Who are the guests summoned to the banquet?

If the wasps flew away, carrying the dead or sickly grubs with them, and dropped them on the ground round about their home, those banqueters would be, first and foremost, the insect-eating birds, the warblers, all of whom are lovers of small game. In this connection, we will allow ourselves a brief digression. We all know with what jealous intolerance the nightingales occupy each his own cantonment. Neighborly intercourse among them is tabooed. The males frequently exchange defiant couplets at a distance; but, should the challenged party draw near, the challenger makes him clear off. Now, not far from my house, in a scanty clump of holly oaks which would barely give the woodcutter the wherewithal for a dozen faggots, I used, all through the spring, to hear such full-throated warbling of nightingales that the songs of those virtuosi, all giving voice at once and with no attempt at order, degenerated into a deafening hubbub.

Why did those passionate devotees of solitude come and settle in such large numbers at a spot where custom decrees that there is just room enough for one household only? What reasons have made the recluse become a congregation? I asked the owner of the spinney about the matter.

'It's like that every year,' he said. 'The clump is overrun by Nightingales.'

'And the reason?'

'The reason is that there is a hive close by, behind that wall.'

I looked at the man in amazement, unable to understand what connection there could be between a hive and the thronging nightingales.

'Why, yes,' he added, 'there are a lot of nightingales because there are a lot of bees.

Another questioning look from my side. I did not yet understand. The explanation came: 'The bees,' he said, 'throw out their dead grubs. The front of the hive is strewn with them in the mornings; and the nightingales come and collect them for themselves and their families. They are very fond of them.'

This time I had solved the puzzle. Delicious food, abundant and fresh each day, had brought the songsters together. Contrary to their habit, numbers of nightingales are living on friendly terms in a cluster of bushes, in order to be near the hive and to have a larger share in the morning distribution of plump dainties.

In the same way, the nightingale and his gastronomical rivals would haunt the neighborhood of the wasps' nests, if the dead grubs were cast out on the surface of the soil; but these delicacies fall inside the burrow and no little bird would dare to enter the murky cave, even if the entrance were not too small to admit it. Other consumers are needed here, small in size and great in daring; the fly is called for and her maggot, the king of the departed. What the greenbottles, the bluebottles and the flesh flies do in the open air, at the expense of every kind of corpse, other flies, narrowing their province, do underground at the Wasps' expense.

Let us turn our attention, in September, to the wrapper of a wasps' nest. On the outer surface and there alone, this wrapper is strewn with a multitude of big, white, elliptical dots, firmly fixed to the brown paper and measuring about two millimeters and a half long by one and a half wide. Flat below, convex above and of a lustrous white, these dots resemble very neat drops fallen from a tallow candle. Lastly, their backs are streaked with faint transversal lines, an elegant detail perceptible only with the lens. These curious objects are scattered all over the surface of the wrapper, sometimes at a distance from one another, sometimes gathered into more or less dense groups. They are the eggs of the Volucella, or bumblebee fly (Volucella zonaria, LIN.)

Also stuck to the brown paper of the outer wrapper and mixed up with the Volucella's are a large number of other eggs, chalk white, spear-shaped and ridged lengthwise with seven or eight thin ribs, after the manner of the seeds of certain Umbelliferae. The finishing touch to their delicate beauty is the fine stippling all over the surface. They are smaller by half than the others. I have seen grubs come out of them which might easily be the earliest stage of some pointed maggots which I have already noticed in the burrows. My attempts to rear them failed; and I am not able to say which fly these eggs belong to. Enough for us to note the nameless one in passing. There are plenty of others, which we must make up our minds to leave unlabelled, in view of the jumbled crowd of feasters in the ruined wasps' nest. We will concern ourselves only with the most remarkable, in the front rank of which stands the bumblebee Fly.

She is a gorgeous and powerful fly; and her costume, with its brown and yellow bands, shows a vague resemblance to that of the wasps. Our fashionable theorists have availed themselves of this brown and yellow to cite the Volucella as a striking instance of protective mimicry. Obliged, if not on her own behalf, at least on that of her family, to introduce herself as a parasite into the wasp's home, she resorts, they tell us, to trickery and craftily dons her victim's livery. Once inside the wasps' nest, she is taken for one of the inhabitants and attends quietly to her business.

The simplicity of the wasp, duped by a very clumsy imitation of her garb, and the depravity of the fly, concealing her identity under a counterfeit presentment, exceed the limits of my credulity. The wasp is not so silly nor the Volucella so clever as we are assured. If the latter really meant to deceive the Wasp by her appearance, we must admit that her disguise is none too successful. Yellow sashes round the abdomen do not make a wasp. It would need more than that and, above all, a slender figure and a nimble carriage; and the Volucella is thickset and corpulent and sedate in her movements. Never will the wasp take that unwieldy insect for one of her own kind. The difference is too great.

Poor Volucella, mimesis has not taught you enough. You ought—this is the essential point—to have adopted a wasp's shape; and that you forgot to do: you remained a fat fly, easily recognizable. Nevertheless, you penetrate into the terrible cavern; you are able to stay there for a long time, without danger, as the eggs profusely strewn on the wrapper of the wasps' nest show. How do you set about it?

Let us, first of all, remember that the bumblebee fly does not enter the enclosure in which the combs are heaped: she keeps to the outer surface of the paper rampart and there lays her eggs. Let us, on the other hand, recall the Polistes [a tree nesting wasp] placed in the company of the wasps in my vivarium. Here of a surety is one who need not have recourse to mimicry to find acceptance. She belongs to the guild, she is a wasp herself. Any of us that had not the trained eye of the entomologist would confuse the two species. Well, this stranger, as long as she does not become too importunate, is quite readily tolerated by the caged wasps. None seeks to pick a quarrel with her. She is even admitted to the table, the strip of paper smeared with honey. But she is doomed if she inadvertently sets foot upon the combs. Her costume, her shape, her size, which tally almost exactly with the costume, shape and size of the wasp, do not save her from her fate. She is at once recognized as a stranger and attacked and slaughtered with the same vigor as the larvae of the Hylotoma sawfly and the Saperda beetle, neither of which bears any outward resemblance to the larva of the wasps.

Seeing that identity of shape and costume does not save the Polistes, how will the Volucella fare, with her clumsy imitation? The wasp's eye, which is able to discern the dissimilar in the like, will refuse to be caught. The moment she is recognized, the stranger is killed on the spot. As to that there is not the shadow of a doubt.

In the absence of bumblebee flies at the moment of experimenting, I employ another fly, Milesia fulminans, who, thanks to her slim figure and her handsome yellow bands, presents a much more striking likeness to the wasp than does the fat Volucella zonaria. Despite this resemblance, if she rashly venture on the combs, she is stabbed and slain. Her yellow sashes, her slender abdomen deceive nobody. The stranger is recognized behind the features of a double.

My experiments under glass, which varied according to the captures which I happened to make, all lead me to this conclusion: as long as there is more propinquity, even around the honey, the other occupants are tolerated fairly well; but, if they touch the cells, they are assaulted and often killed, without distinction of shape or costume. The grubs' dormitory is the sanctum sanctorum which no outsider must enter under pain of death.

With these caged captives I experiment by daylight, whereas the free wasps work in the absolute darkness of their underground retreat. Where light is absent, color goes for nothing. Once, therefore, that she has entered the cavern, the bumblebee fly derives no benefit from her yellow bands, which are supposed to be her safeguard. Whether garbed as she is or otherwise, it is easy for her to effect her purpose in the dark, on condition that she avoids the tumultuous interior of the wasps' nest. So long as she has the prudence not to hustle the passers by, she can dab her eggs, without danger, on the paper wall. No one will know of her presence. The dangerous thing is to cross the threshold of the burrow in broad daylight, before the eyes of those who go in and out. At that moment alone, protective mimicry would be convenient. Now does the entrance of the Volucella into the presence of a few wasps entail such very great risks? The wasps' nest in my enclosure, the one which was afterwards to perish in the sun under a bell glass, gave me the opportunity for prolonged observations, but without any result upon the subject of my immediate concern. The bumblebee fly did not appear. The period for her visits had doubtless passed; for I found plenty of her grubs when the nest was dug up.

Other flies rewarded me for my assiduity. I saw some—at a respectful distance, I need hardly say—entering the burrow. They were insignificant in size and of a dark gray color, not unlike that of the housefly. They had not a patch of yellow about them and certainly had no claim to protective mimicry. Nevertheless, they went in and out as they pleased, calmly, as though they were at home. As long as there was not too great a number at the door, the wasps left them alone. When there was anything of a crowd, the gray visitors waited near the threshold for a less busy moment. No harm came to them.

Inside the establishment, the same peaceful relations prevail. In this respect I have the evidence of my excavations. In the underground charnel house, so rich in Fly grubs, I find no corpses of adult flies. If the strangers had been slaughtered in passing through the entrance hall, or lower down, they would fall to the bottom of the burrow anyhow, with the other rubbish. Now in this charnel house, as I said, there are never any dead bumblebee flies, never a fly of any sort. The incomers are respected. Having done their business, they go out unscathed.

This tolerance on the part of the wasps is surprising. And a suspicion comes to one's mind: can it be that the Volucella and the rest are not what the accepted theories of natural history call them, namely, enemies, grub killers sacking the wasps' nest? We will look into this by examining them when they are hatched. Nothing is easier, in September and October, than to collect the Volucella's eggs in such numbers as we please. They abound on the outer surface of the wasps' nest. Moreover, as with the larvae of the wasp, it is some time before they are suffocated by the petroleum fumes; and so most of them are sure to hatch. I take my scissors, cut the most densely populated bits from the paper wall of the nest and fill a jar with them. This is the warehouse from which I shall daily, for the best part of the next two months, draw my supply of nascent grubs.

The Volucella's egg remains where it is, with its white color always strongly marked against the brown of the background. The shell wrinkles and collapses; and the fore end tears open. From it there issues a pretty little white grub, thin in front, swelling slightly in the rear and bristling all over with fleshy protuberances. The creature's papillae are set on its sides like the teeth of a comb; at the rear, they lengthen and spread into a fan; on the back, they are shorter and arranged in four longitudinal rows. The last section but one carries two short, bright red breathing tubes, standing aslant and joined to each other. The fore part, near the pointed mouth, is of a darker, brownish color. This is the biting and motor apparatus, seen through the skin and consisting of two fangs. Taken all round, the grub is a pretty little thing, with its bristling whiteness, which gives it the appearance of a tiny snowflake. But this elegance does not last long: grown big and strong, the bumblebee fly's grub becomes soiled with sanies, turns a russety brown and crawls about in the guise of a hulking porcupine.

What becomes of it when it leaves the egg? This my warehousing jar tells me, partly. Unable to keep its balance on sloping surfaces, it drops to the bottom of the receptacle, where I find it, daily, as hatched, wandering restlessly. Things must happen likewise at the wasps'. Incapable of standing on the slant of the paper wall, the newborn grubs slide to the bottom of the underground cavity, which contains, especially at the end of the summer, a heaped up provender of deceased wasps and dead larvae removed from the cells and flung outside the house, all nice and gamy, as proper maggot's food should be. The Volucella's offspring, themselves maggots, notwithstanding their snowy apparel, find in this charnel house victuals to their liking, incessantly renewed. Their fall from the high walls might well be not accidental, but rather a means of reaching, quickly and without searching, the good things down at the bottom of the cavern. Perhaps, also, some of the white grubs, thanks to the holes that make the wrapper resemble a spongy cover, manage to slip inside the Wasps' nest. Still, most of the Volucella's grubs, at whatever stage of their development, are in the basement of the burrow, among the carrion remains. The others, those settled in the wasps' home itself, are comparatively few.

These returns are enough to show us that the grubs of the bumblebee fly do not deserve the bad reputation that has been given them. Satisfied with the spoils of the dead, they do not touch the living; they do not ravage the wasps' nest: they disinfect it.

Experiment confirms what we have learnt in the actual nests. Over and over again, I bring wasp grubs and Volucella grubs together in small test tubes, which are easy to observe. The first are well and strong; I have just taken them from their cells. The others are in various stages, from that of the snowflake born the same day to that of the sturdy porcupine. There is nothing tragic about the encounter. The grubs of the bumblebee fly roam about the test-tube without touching the live tidbit. The most that they do is to put their mouths for a moment to the morsel; then they take it away again, not caring for the dish.

They want something different: a wounded, a dying grub; a corpse dissolving into sanies. Indeed, if I prick the wasp grub with a needle, the scornful ones at once come and sup at the bleeding wound. If I give them a dead grub, brown with putrefaction, the worms rip it open and feast on its humors. Better still: I can feed them quite satisfactorily with wasps that have turned putrid under their horny rings; I see them greedily suck the juices of decomposing Rosechafer grubs; I can keep them thriving with chopped up butcher's meat, which they know how to liquefy by the method of the common maggot. And these unprejudiced ones, who accept anything that comes their way, provided it be dead, refuse it when it is alive. Like the true flies that they are, frank body snatchers, they wait, before touching a morsel, for death to do its work.

Inside the wasps' nest, robust grubs are the rule and weaklings the rare exception, because of the assiduous supervision which eliminates anything that is diseased and like to die. Here, nevertheless, Volucella grubs are found, on the combs, among the busy wasps. They are not, it is true, so numerous as in the charnel house below, but still pretty frequent. Now what do they do in this abode where there are no corpses? Do they attack the healthy? Their continual visits from cell to cell would at first make one think so; but we shall soon be undeceived if we observe their movements closely; and this is possible with my glass roofed colonies.

I see them fussily crawling on the surface of the combs, curving their necks from side to side and taking stock of the cells. This one does not suit, nor that one either; the bristly creature passes on, still in search, thrusting its pointed fore part now here, now there. This time, the cell appears to fulfil the requisite conditions. A larva, glowing with health, opens wide its mouth, believing its nurse to be approaching. It fills the hexagonal chamber with its bulging sides.

The gluttonous visitor bends and slides its slender fore part, a blade of exquisite suppleness, between the wall and the inhabitant, whose slack rotundity yields to the pressure of this animated wedge. It plunges into the cell, leaving no part of itself outside but its wide hind quarters, with the red dots of the two breathing tubes.

It remains in this posture for some time, occupied with its work at the bottom of the cell. Meanwhile, the wasps present do not interfere, remain impassive, showing that the grub visited is in no peril. The stranger, in fact, withdraws with a soft, gliding motion. The chubby babe, a sort of India rubber bag, resumes its original volume without having suffered any harm, as its appetite proves. A nurse offers it a mouthful, which it accepts with every sign of unimpaired vigor. As for the Volucella grub, it licks its lips after its own fashion, pushing its two fangs in and out; then, without further loss of time, goes and repeats its probing elsewhere.

What it wants down there, at the bottom of the cells, behind the grubs, cannot be decided by direct observation; it must be guessed at. Since the visited larva remains intact, it is not prey that the Volucella grub is after. Besides, if murder formed part of its plans, why descend to the bottom of the cell, instead of attacking the defenseless recluse straight way? It would be much easier to suck the patient's juices through the actual orifice of the cell. Instead of that, we see a dip, always a dip and never any other tactics.

Then what is there behind the wasp grub? Let us try to put it as decently as possible. In spite of its exceeding cleanliness, this grub is not exempt from the physiological ills inseparable from the stomach. Like all that eats, it has intestinal waste matter with regard to which its confinement compels it to behave with extreme discretion. Like so many other close-cabined larvae of Wasps and Bees, it waits until the moment of the transformation to rid itself of its digestive refuse. Then, once and for all, it casts out the unclean accumulation whereof the pupa, that delicate, reborn organism, must not retain the least trace. This is found later, in any empty cell, in the form of a dark purple plug. But, without waiting for this final purge, this lump, there are, from time to time, slight excretions of fluid, clear as water. We have only to keep a Wasp grub in a little glass tube to recognize these occasional discharges. Well, I see nothing else to explain the action of the Volucella's grubs when they dip into the cells without wounding the larvae. They are looking for this liquid, they provoke its emission. It represents to them a dainty which they enjoy over and above the more substantial fare provided by the corpses.

The bumblebee fly, that sanitary inspector of the Vespine city, fulfils a double office: she wipes the wasp's children and she rids the nest of its dead. For this reason, she is peacefully received, as an auxiliary, when she enters the burrow to lay her eggs; for this reason, her grub is tolerated, nay more, respected, in the very heart of the dwelling, where none might stray with impunity. I remember the brutal reception given to the Saperda and Hylotoma grubs when I place them on a comb. Forthwith grabbed, bruised and riddled with stings, the poor wretches perish. It is quite a different matter with the offspring of the Volucella. They come and go as they please, poke about in the cells, elbow the inhabitants and remain unmolested. Let us give some instances of this clemency, which is very strange in the irascible Wasp.

For a couple of hours, I fix my attention on a Volucella grub established in a cell, side by side with the Wasp grub, the mistress of the house. The hind quarters emerge, displaying their papillae. Sometimes also the fore part, the head, shows, bending from side to side with sudden, snake-like motions. The wasps have just filled their crops at the honey pot; they are dispensing the rations, are very busily at work; and things are taking place in broad daylight, on the table by the window.

As they pass from cell to cell, the nurses repeatedly brush against and stride across the Volucella grub. There is no doubt that they see it. The intruder does not budge, or, if trodden on, curls up, only to reappear the next moment. Some of the wasps stop, bend their heads over the opening, seem to be making inquiries and then go off, without troubling further about the state of things. One of them does something even more remarkable: she tries to give a mouthful to the lawful occupant of the cell; but the larva, which is being squeezed by its visitor, has no appetite and refuses. Without the least sign of anxiety on behalf of the nursling which she sees in awkward company, the wasp retires and goes to distribute its ration elsewhere. In vain I prolong my examination: there is no fluster of any kind. The Volucella grub is treated as a friend, or at least as a visitor that does not matter. There is no attempt to dislodge it, to worry it, to put it to flight. Nor does the grub seem to trouble greatly about those who come and go. Its tranquillity, tells us that it feels at home.

Here is some further evidence: the grub has plunged, head downwards, into an empty cell, which is too small to contain the whole of it. Its hindquarters stick out, very visibly. For long hours, it remains motionless in this position. At every moment, wasps pass and repass close by. Three of them, at one time together, at another separately, come and nibble at the edges of the cell; they break off particles which they reduce to paste for a new piece of work. The passers by, intent upon their business, may not perceive the intruder; but these three certainly do. During their work of demolition, they touch the grub with their legs, their antennae, their palpi; and yet none of them minds it. The fat grub, so easily recognized by its queer figure, is left alone; and this in broad daylight, where everybody can see it. What must it be when the profound darkness of the burrows protects the visitor with its mysteries!

I have been experimenting all along with big Volucella grubs, colored with the dirty red which comes with age. What effect will pure white produce? I sprinkle on the surface of the combs some larvae that have lately left the egg. The tiny, snow-white grubs make for the nearest cells, go down into them, come out again and hunt elsewhere. The wasps peaceably let them go their way, as heedless of the little white invaders as of the big red ones. Sometimes, when it enters an occupied cell, the little creature is seized by the owner, the wasp grub, which nabs it and turns and returns it between its mandibles. Is this a defensive bite? No, the wasp grub has merely blundered, taking its visitor for a proffered mouthful. There is no great harm done. Thanks to its suppleness, the little grub emerges from the grip intact and continues its investigations.

It might occur to us to attribute this tolerance to some lack of penetration in the wasps' vision. What follows will undeceive us: I place separately, in empty cells, a grub of Saperda scalaria and a Volucella grub, both of them white and selected so as not to fill the cell entirely. Their presence is revealed only by the paleness of the hind part which serves as a plug to the opening. A superficial examination would leave the nature of the recluse undecided. The wasps make no mistake: they extirpate the Saperda grub, kill it, fling it on the dust heap; they leave the Volucella grub in peace.

The two strangers are quite well recognized in the secrecy of the cells: one is the intruder that must be turned out; the other is the regular visitor that must be respected. Sight helps, for things take place in the daylight, under glass; but the wasps have other means of information in the dimness of the burrow. When I produce darkness by covering the apparatus with a screen, the murder of the trespassers is accomplished just the same. For so say the police regulations of the wasps' nest: any stranger discovered must be slain and thrown on the midden.

To thwart this vigilance, the real enemies need to be masters of the art of deceptive immobility and cunning disguise. But there is no deception about the Volucella grub. It comes and goes, openly, wheresoever it will; it looks round amongst the wasps for cells to suit it. What has it to make itself thus respected? Strength? Certainly not. It is a harmless creature, which the wasp could rip open with a blow of her shears, while a touch of the sting would mean lightning death. It is a familiar guest, to whom no denizen of a wasps' nest bears any ill will. Why? Because it renders good service: so far from working mischief, it does the scavenging for its hosts. Were it an enemy or merely an intruder, it would be exterminated; as a deserving assistant, it is respected.

Then what need is there for the Volucella to disguise herself as a wasp? Any fly, whether clad in drab or motley, is admitted to the burrow directly she makes herself useful to the community. The mimicry of the bumblebee fly, which was said to be one of the most conclusive cases, is, after all, a mere childish notion. Patient observation, continually face to face with facts, will have none of it and leaves it to the armchair naturalists, who are too prone to look at the animal world through the illusive mists of theory.

CHAPTER XII.
MATHEMATICAL MEMORIES: NEWTON'S BINOMIAL THEOREM

The spider's web is a glorious mathematical problem. I should enjoy working it out in all its details, were I not afraid of wearying the reader's attention. Perhaps I have even gone too far in the little that I have said, in which case I owe him some compensation: 'Would you like me,' I will ask him, 'would you like me to tell you how I acquired sufficient algebra to master the logarithmic systems and how I became a surveyor of Spiders' webs? Would you? It will give us a rest from natural history.'

I seem to catch a sign of acquiescence. The story of my village school, visited by the chicks and the porkers, has been received with some indulgence; why should not my harsh school of solitude possess its interest as well? Let us try to describe it. And who knows? Perhaps, in doing so, I shall revive the courage of some other poor derelict hungering after knowledge.

I was denied the privilege of learning with a master. I should be wrong to complain. Solitary study has its advantages: it does not cast you in the official mould; it leaves you all your originality. Wild fruit, when it ripens, has a different taste from hothouse produce: it leaves on a discriminating palate a bittersweet flavor whose virtue is all the greater for the contrast. Yes, if it were in my power, I would start afresh, face to face with my only counselor, the book itself, not always a very lucid one; I would gladly resume my lonely watches, my struggles with the darkness whence, at last, a glimmer appears as I continue to explore it; I should retraverse the irksome stages of yore, stimulated by the one desire that has never failed me, the desire of learning and of afterwards bestowing my mite of knowledge on others.

When I left the normal school, my stock of mathematics was of the scantiest. How to extract a square root, how to calculate and prove the surface of a sphere: these represented to me the culminating points of the subject. Those terrible logarithms, when I happened to open a table of them, made my head swim, with their columns of figures; actual fright, not unmixed with respect, overwhelmed me on the very threshold of that arithmetical cave. Of algebra I had no knowledge whatever. I had heard the name; and the syllables represented to my poor brain the whole whirling legion of the abstruse.

Besides, I felt no inclination to decipher the alarming hieroglyphics. They made one of those indigestible dishes which we confidently extol without touching them. I greatly preferred a fine line of Virgil, whom I was now beginning to understand; and I should have been surprised indeed had any one told me that, for long years to come, I should be an enthusiastic student of the formidable science. Good fortune procured me my first lesson in algebra, a lesson given and not received, of course.

A young man of about my own age came to me and asked me to teach him algebra. He was preparing for his examination as a civil engineer; and he came to me because, ingenuous youth that he was, he took me for a well of learning. The guileless applicant was very far out in his reckoning.

His request gave me a shock of surprise, which was forthwith repressed on reflection: 'I give algebra lessons?' said I to myself. 'It would be madness: I don't know anything about the subject!'

And I left it at that for a moment or two, thinking hard, drawn now this way, now that with indecision: 'Shall I accept? Shall I refuse?' continued the inner voice.

Pooh, let's accept! An heroic method of learning to swim is to leap boldly into the sea. Let us hurl ourselves head first into the algebraical gulf; and perhaps the imminent danger of drowning will call forth efforts capable of bringing me to land. I know nothing of what he wants. It makes no difference: let's go ahead and plunge into the mystery. I shall learn by teaching.

It was a fine courage that drove me full tilt into a province which I had not yet thought of entering. My twenty-year-old confidence was an incomparable lever.

'Very well,' I replied. 'Come the day after tomorrow, at five, and we'll begin.'

This twenty-four hours' delay concealed a plan. It secured me the respite of a day, the blessed Thursday, which would give me time to collect my forces.

Thursday comes. The sky is gray and cold. In this horrid weather, a grate well filled with coke has its charms. Let's warm ourselves and think.

Well, my boy, you've landed yourself in a nice predicament! How will you manage tomorrow? With a book, plodding all through the night, if necessary, you might scrape up something resembling a lesson, just enough to fill the dread hour more or less. Then you could see about the next: sufficient for the day is the evil thereof. But you haven't the book. And it's no use running out to the bookshop. Algebraical treatises are not current wares. You'll have to send for one, which will take a fortnight at least. And I've promised for tomorrow, for tomorrow certain! Another argument and one that admits of no reply: funds are low; my last pecuniary resources lie in the corner of a drawer. I count the money: it amounts to twelve sous, which is not enough.

Must I cry off? Rather not! One resource suggests itself: a highly improper one, I admit, not far removed indeed from larceny. O quiet paths of algebra, you are my excuse for this venial sin! Let me confess the temporary embezzlement.

Life at my college is more or less cloistered. In return for a modest payment, most of us masters are lodged in the building; and we take our meals at the principal's table. The science master, who is the big gun of the staff and lives in the town, has nevertheless, like ourselves, his own two cells, in addition to a balcony, or leads, where the chemical preparations give forth their suffocating gases in the open air. For this reason, he finds it more convenient to hold his class here during the greater part of the year. The boys come to these rooms in winter, in front of a grate stuffed full of coke, like mine, and there find a blackboard, a pneumatic trough, a mantelpiece covered with glass receivers, panoplies of bent tubes on the walls, and, lastly, a certain cupboard in which I remember seeing a row of books, the oracles consulted by the master in the course of his lessons.

'Among those books,' said I to myself, 'there is sure to be one on algebra. To ask the owner for the loan of it does not appeal to me. My amiable colleague would receive me superciliously and laugh at my ambitious aims. I am sure he would refuse my request.'

The future was to show that my distrust was justified. Narrow mindedness and petty jealousy prevail everywhere alike.

I decide to help myself to this book, which I should never get by asking. This is the half-holiday. The science master will not put in an appearance today; and the key of my room is practically the same as his. I go, with eyes and ears on the alert. My key does not quite fit; it sticks a little, then goes in; and an extra effort makes it turn in the lock. The door opens. I inspect the cupboard and find that it does contain an algebra book, one of the big, fat books which men used to write in those days, a book nearly half a foot thick. My legs give way beneath me. You poor specimen of a housebreaker, suppose you were caught at it! However, all goes well. Quick, let's lock the door again and go back to our own quarters with the pilfered volume.

And now we are together, O mysterious tome, whose Arab name breathes a strange mustiness of occult lore and claims kindred with the sciences of almagest and alchemy. What will you show me? Let us turn the leaves at random. Before fixing one's eyes on a definite point in the landscape, it is well to take a summary view of the whole. Page follows swiftly upon page, telling me nothing. A chapter catches my attention in the middle of the volume; it is headed, Newton's Binomial Theorem.

The title allures me. What can a binomial theorem be, especially one whose author is Newton, the great English mathematician who weighed the worlds? What has the mechanism of the sky to do with this? Let us read and seek for enlightenment. With my elbows on the table and my thumbs behind my ears, I concentrate all my attention.

I am seized with astonishment, for I understand! There are a certain number of letters, general symbols which are grouped in all manner of ways, taking their places here, there and elsewhere by turns; there are, as the text tells me, arrangements, permutations and combinations. Pen in hand, I arrange, permute and combine. It is a very diverting exercise, upon my word, a game in which the test of the written result confirms the anticipations of logic and supplements the shortcomings of one's thinking apparatus.

'It will be plain sailing,' said I to myself, 'if algebra is no more difficult than this.'

I was to recover from the illusion later, when the binomial theorem, that light, crisp biscuit, was followed by heavier and less digestible fare. But, for the moment, I had no foretaste of the future difficulties, of the pitfall in which one becomes more and more entangled, the longer one persists in struggling. What a delightful afternoon that was, before my grate, amid my permutations and combinations! By the evening, I had nearly mastered my subject. When the bell rang, at seven, to summon us to the common meal at the principal's table, I went downstairs puffed up with the joys of the newly initiated neophyte. I was escorted on my way by a, b and c, intertwined in cunning garlands.

Next day, my pupil is there. Blackboard and chalk, everything is ready. Not quite so ready is the master. I bravely broach my binomial theorem. My hearer becomes interested in the combinations of letters. Not for a moment does he suspect that I am putting the cart before the horse and beginning where we ought to have finished. I relieve the dryness of my explanations with a few little problems, so many halts at which the mind takes breath awhile and gathers strength for fresh flights.

We try together. Discreetly, so as to leave him the merit of the discovery, I shed a little light on the path. The solution is found. My pupil triumphs; so do I, but silently, in my inner consciousness, which says:

'You understand, because you succeed in making another understand.'

The hour passed quickly and very pleasantly for both of us. My young man was contented when he left me; and I no less so, for I perceived a new and original way of learning things.

The ingenious and easy arrangement of the binomial gave me time to tackle my algebra book from the proper commencement. In three or four days, I had rubbed up my weapons. There was nothing to be said about addition and subtraction: they were so simple as to force themselves upon one at first sight. Multiplication spoilt things. There was a certain rule of signs which declared that minus multiplied by minus made plus. How I toiled over that wretched paradox! It would seem that the book did not explain this subject clearly, or rather employed too abstract a method. I read, reread and meditated in vain: the obscure text retained all its obscurity. That is the drawback of books in general: they tell you what is printed in them and nothing more. If you fail to understand, they never advise you, never suggest an attempt along another road which might lead you to the light. The merest word would sometimes be enough to put you on the right track; and that word the books, hidebound in a regulation phraseology, never give you.

How greatly preferable is the oral lesson! It goes forward, goes back, starts afresh, walks around the obstacle and varies the methods of attack until, at long last, light is shed upon the darkness. This incomparable beacon of the master's word was what I lacked; and I went under, without hope of succor, in that treacherous pool of the rule of signs.

My pupil was bound to suffer the effects. After an attempt at an explanation in which I made the most of the few gleams that reached me I asked him:

'Do you understand?'

It was a futile question, but useful for gaining time. Myself not understanding, I was convinced beforehand that he did not understand either.

'No,' he replied, accusing himself, perhaps, in his simple mind, of possessing a brain incapable of taking in those transcendental verities.

'Let us try another method.'

And I start again this way and that way and yet another way. My pupil's eyes serve as my thermometer and tell me of the progress of my efforts. A blink of satisfaction announces my success. I have struck home, I have found the joint in the armor. The product of minus multiplied by minus delivers its mysteries to us.

And thus we continued our studies: he, the passive receiver, taking in the ideas acquired without effort; I, the fierce pioneer, blasting my rock, the book, with the aid of much sitting up at night, to extract the diamond, truth. Another and no less arduous task fell to my share: I had to cut and polish the recondite gem, to strip it of its ruggedness and present it to my companion's intelligence under a less forbidding aspect. This diamond cutter's work, which admitted a little light into the precious stone, was the favorite occupation of my leisure; and I owe a great deal to it.

The ultimate result was that my pupil passed his examination. As for the book borrowed by stealth, I restored it to the shelves and replaced it by another, which, this time, belonged to me.

At my normal school, I had learnt a little elementary geometry under a master. From the first few lessons onwards, I rather enjoyed the subject. I divined in it a guide for one's reasoning faculties through the thickets of the imagination; I caught a glimpse of a search after truth that did not involve too much stumbling on the way, because each step forward rests solidly upon the step already taken; I suspected geometry to be what it preeminently is: a school of intellectual fencing.

The truth demonstrated and its application matter little to me; what rouses my enthusiasm is the process that sets the truth before us. We start from a brilliantly lighted spot and gradually get deeper and deeper in the darkness, which, in its turn, becomes self-illuminated by kindling new lights for a higher ascent. This progressive march of the known toward the unknown, this conscientious lantern lighting what follows by the rays of what comes before: that was my real business.

Geometry was to teach me the logical progression of thought; it was to tell me how the difficulties are broken up into sections which, elucidated consecutively, together form a lever capable of moving the block that resists any direct efforts; lastly, it showed me how order is engendered, order, the base of clarity. If it has ever fallen to my lot to write a page or two which the reader has run over without excessive fatigue, I owe it, in great part, to geometry, that wonderful teacher of the art of directing one's thought. True, it does not bestow imagination, a delicate flower blossoming none knows how and unable to thrive on every soil; but it arranges what is confused, thins out the dense, calms the tumultuous, filters the muddy and gives lucidity, a superior product to all the tropes of rhetoric.

Yes, as a toiler with the pen, I owe much to it. Wherefore my thoughts readily turn back to those bright hours of my novitiate, when, retiring to a corner of the garden in recreation time, with a bit of paper on my knees and a stump of pencil in my fingers, I used to practice deducing this or that property correctly from an assemblage of straight lines. The others amused themselves all around me; I found my delight in the frustum of a pyramid. Perhaps I should have done better to strengthen the muscles of my thighs by jumping and leaping, to increase the suppleness of my loins with gymnastic contortions. I have known some contortionists who have prospered beyond the thinker.

See me then entering the lists as an instructor of youth, fairly well acquainted with the elements of geometry. In case of need, I could handle the land surveyor's stake and chain. There my views ended. To cube the trunk of a tree, to gauge a cask, to measure the distance of an inaccessible point appeared to me the highest pitch to which geometrical knowledge could hope to soar. Were there loftier flights? I did not even suspect it, when an unexpected glimpse showed me the puny dimensions of the little corner which I had cleared in the measureless domain.

At that time, the college in which, two years before, I had made my first appearance as a teacher, had just halved the size of its classes and largely increased its staff. The newcomers all lived in the building, like myself, and we had our meals in common at the principal's table. We formed a hive where, in our leisure time, some of us, in our respective cells, worked up the honey of algebra and geometry, history and physics, Greek and Latin most of all, sometimes with a view to the class above, sometimes and oftener with a view to acquiring a degree. The university titles lacked variety. All my colleagues were bachelors of letters, but nothing more. They must, if possible, arm themselves a little better to make their way in the world. We all worked hard and steadily. I was the youngest of the industrious community and no less eager than the rest to increase my modest equipment.

Visits between the different rooms were frequent. We would come to consult one another about a difficulty, or simply to pass the time of day. I had as a neighbor, in the next cell to mine, a retired quartermaster who, weary of barrack life, had taken refuge in education. When in charge of the books of his company he had become more or less familiar with figures; and it became his ambition to take a mathematical degree. His cerebrum appears to have hardened while he was with his regiment. According to my dear colleagues, those amiable retailers of the misfortunes of others, he had already twice been plucked. Stubbornly, he returned to his books and exercises, refusing to be daunted by two reverses.

It was not that he was allured by the beauties of mathematics, far from it; but the step to which he aspired favored his plans. He hoped to have his own boarders and dispense butter and vegetables to lucrative purpose. The lover of study for its own sake and the persistent trapper hunting a diploma as he would something to put in his mouth were not made to understand or to see much of each other. Chance, however, brought us together.

I had often surprised our friend sitting in the evening, by the light of a candle, with his elbows on the table and his head between his hands, meditating at great length in front of a big exercise book crammed with cabalistic signs. From time to time, when an idea came to him, he would take his pen and hastily put down a line of writing wherein letters, large and small, were grouped without any grammatical sense. The letters x and y often recurred, intermingled with figures. Every row ended with the sign of equality and a nought. Next came more reflection, with closed eyes, and a fresh row of letters arranged in a different order and likewise followed by a nought. Page after page was filled in this queer fashion, each line winding up with 0.

'What are you doing with all those rows of figures amounting to zero?' I asked him one day.

The mathematician gave me a leery look, picked up in barracks. A sarcastic droop in the corner of his eye showed how he pitied my ignorance. My colleague of the many noughts did not, however, take an unfair advantage of his superiority. He told me that he was working at analytical geometry.

The phrase had a strange effect upon me. I ruminated silently to this purpose: there was a higher geometry, which you learnt more particularly with combinations of letters in which x and y played a prominent part. When my next-door neighbor reflected so long, clutching his forehead between his hands, he was trying to discover the hidden meaning of his own hieroglyphics; he saw the ghostly translation of his sums dancing in space. What did he perceive? How would the alphabetical signs, arranged first in one and then in another manner, give an image of the actual things, an image visible to the eyes of the mind alone? It beat me.

'I shall have to learn analytical geometry some day,' I said. 'Will you help me?'

'I'm quite willing,' he replied, with a smile in which I read his lack of confidence in my determination.

No matter; we struck a bargain that same evening. We would together break up the stubble of algebra and analytical geometry, the foundation of the mathematical degree; we would make common stock: he would bring long hours of calculation, I my youthful ardor. We would begin as soon as I had finished with my arts degree, which was my main preoccupation for the moment.

In those far off days it was the rule to make a little serious literary study take precedence of science. You were expected to be familiar with the great minds of antiquity, to converse with Horace and Virgil, Theocritus and Plato, before touching the poisons of chemistry or the levers of mechanics. The niceties of thought could only be the gainers by these preparations. Life's exigencies, ever harsher as progress afflicts us with its increasing needs, have changed all that. A fig for correct language! Business before all!

This modern hurry would have suited my impatience. I confess that I fumed against the regulation which forced Latin and Greek upon me before allowing me to open up relations with the sine and cosine. Today, wiser, ripened by age and experience, I am of a different opinion. I very much regret that my modest literary studies were not more carefully conducted and further prolonged. To fill up this enormous blank a little, I respectfully returned, somewhat late in life, to those good old books which are usually sold second-hand with their leaves hardly cut. Venerable pages, annotated in pencil during the long evenings of my youth, I have found you again and you are more than ever my friends. You have taught me that an obligation rests upon whoever wields the pen: he must have something to say that is capable of interesting us. When the subject comes within the scope of natural science, the interest is nearly always assured; the difficulty, the great difficulty, is to prune it of its thorns and to present it under a prepossessing aspect. Truth, they say, rises naked from a well. Agreed; but admit that she is all the better for being decently clothed. She craves, if not the gaudy furbelows borrowed from rhetoric's wardrobe, at least a vine leaf. The geometers alone have the right to refuse her that modest garment; in theorems, plainness suffices. The others, especially the naturalist, are in duty bound to drape a gauze tunic more or less elegantly around her waist.

Suppose I say: 'Baptiste, give me my slippers.'

I am expressing myself in plain language, a little poor in variants. I know exactly what I am saying and my speech is understood.

Others—and they are numerous—contend that this rudimentary method is the best in all things. They talk science to their readers as they might talk slippers to Baptiste. Kaffir syntax does not shock them. Do not speak to them of the value of a well selected term, set down in its right place, still less of a lilting construction, sounding rather well. Childish nonsense they call all that; the fiddling of a short sighted mind!

Perhaps they are right: the Baptiste idiom is a great economizer of time and trouble. This advantage does not tempt me; it seems to me that an idea stands out better if expressed in lucid language, with sober imagery. A suitable phrase, placed in its correct position and saying without fuss the things we want to say, necessitates a choice, an often laborious choice.

There are drab words, the commonplaces of colloquial speech; and there are, so to speak, colored words, which may be compared with the brushstrokes strewing patches of light over the gray background of a painting. How are we to find those picturesque words, those striking features which arrest the attention? How are we to group them into a language heedful of syntax and not displeasing to the ear?

I was taught nothing of this art. For that matter, is it ever taught in the schools? I greatly doubt it. If the fire that runs through our veins, if inspiration do not come to our aid, we shall flutter the pages of the thesaurus in vain: the word for which we seek will refuse to come. Then to what masters shall we have recourse to quicken and develop the humble germ that is latent within us? To books.

As a boy, I was always an ardent reader; but the niceties of a well-balanced style hardly interested me: I did not understand them. A good deal later, when close upon fifteen, I began vaguely to see that words have a physiognomy of their own. Some pleased me better than others by the distinctness of their meaning and the resonance of their rhythm; they produced a clearer image in my mind; after their fashion, they gave me a picture of the object described. Colored by its adjective and vivified by its verb, the name became a living reality: what it said I saw. And thus, gradually, was the magic of words revealed to me, when the chances of, my undirected reading placed a few easy standard pages in my way.

CHAPTER XIII. MATHEMATICAL MEMORIES: MY LITTLE TABLE

It is time to start our analytical geometry. He can come now, my partner, the mathematician: I think I shall understand what he says. I have already run through my book and noticed that our subject, whose beautiful precision makes work a recreation, bristles with no very serious difficulties.

We begin in my room, in front of a blackboard. After a few evenings, prolonged into the peaceful watches of the night, I become aware, to my great surprise, that my teacher, the past master in those hieroglyphics, is really, more often than not, my pupil. He does not see the combinations of the abscissas and ordinates very clearly. I make bold to take the chalk in hand myself, to seize the rudder of our algebraical boat. I comment on the book, interpret it in my own fashion, expound the text, sound the reefs until daylight comes and leads us to the haven of the solution. Besides, the logic is so irresistible, it is all such easy going and so lucid that often one seems to be remembering rather than learning.

And so we proceed, with our positions reversed. I dig into the hard rock, crumble it, loosen it until I make room for thought to penetrate. My comrade—I can now allow myself to speak of him on equal terms—my comrade listens, suggests objections, raises difficulties which we try to solve in unison. The two combined levers, inserted in the fissure, end by shaking and overturning the rocky mass.

I no longer see in the corner of the quartermaster's eye the leery droop that greeted me at the start. Cordial frankness now reigns, the infectious high spirits imparted by success. Little by little, dawn breaks, very misty as yet, but laden with promises. We are both greatly amazed; and my share in the satisfaction is a double one, for he sees twice over who makes others see. Thus do we pass half the night, in delightful hours. We cease when sleep begins to weigh too heavily on our eyelids.

When my comrade returns to his room, does he sleep, careless for the moment of the shifting scene which we have conjured up? He confesses to me that he sleeps soundly. This advantage I do not possess. It is not in my power to pass the sponge over my poor brain even as I pass it over the blackboard. The network of ideas remains and forms as it were a moving cobweb in which repose wriggles and tosses, incapable of finding a stable equilibrium. When sleep does come at last, it is often but a state of somnolence which, far from suspending the activity of the mind, actually maintains and quickens it more than waking would. During this torpor, in which night has not yet closed upon the brain, I sometimes solve mathematical difficulties with which I struggled unsuccessfully the day before. A brilliant beacon, of which I am hardly conscious, flares in my brain. Then I jump out of bed, light my lamp again and hasten to jot down my solutions, the recollection of which I should have lost on awakening. Like lightning flashes, those gleams vanish as suddenly as they appear.

Whence do they come? Probably from a habit which I acquired very early in life: to have food always there for my mind, to pour the never failing oil constantly into the lamp of thought. Would you succeed in the things of the mind? The infallible method is to be always thinking of them. This method I practiced more sedulously than my comrade; and hence, no doubt, arose the interchange of positions, the disciple turned into the master. It was not, however, an overwhelming infatuation, a painful obsession; it was rather a recreation, almost a poetic feast. As our great lyric writer put it in the preface to his volume, Les Rayons et les ombres: 'Mathematics play their part in art as well as in science. There is algebra in astronomy: astronomy is akin to poetry; there is algebra in music: music is akin to poetry.'

Is this poetic exaggeration? Surely not: Victor Hugo spoke truly. Algebra, the poem of order, has magnificent flights. I look upon its formulae, its strophes as superb, without feeling at all astonished when others do not agree. My colleague's satirical look came back when I was imprudent enough to confide my extrageometrical raptures to his ears: 'Nonsense,' said he, 'pure stuff and nonsense! Let's get on with our tangents.'

The quartermaster was right: the strict severity of our approaching examination allowed of no such dreamer's outbursts. Was I, on my side, very wrong? To warm chill calculation by the fire of the ideal, to lift one's thought above mere formulae, to brighten the caverns of the abstract with a spark of life: was this not to ease the effort of penetrating the unknown? Where my comrade plodded on, scorning my viaticum, I performed a journey of pleasure. If I had to lean on the rude staff of algebra, I had for my guide that voice within me, urging me to lofty flights. Study became a joy.

It became still more interesting when, after the angularities of a combination of straight lines, I learnt to portray the graces of a curve. How many properties were there of which the compass knew nothing, how many cunning laws lay contained in embryo within an equation, the mysterious nut which must be artistically cracked to extract the rich kernel, the theorem! Take this or that term, place the + sign before it and forthwith you have the ellipse, the trajectory of the planets, with its two friendly foci, transmitting pairs of vectors whose sum is constant; substitute the—sign and you have the hyperbola with the antagonistic foci, the desperate curve that dives into space with infinite tentacles, approaching nearer and nearer to straight lines, the asymptotes, but never succeeding in meeting them. Suppress that term and you have the parabola, which vainly seeks in infinity its lost second focus; you have the trajectory of the bombshell; you have the path of certain comets which come one day to visit our sun and then flee to depths whence they never return. Is it not wonderful thus to formulate the orbit of the worlds? I thought so then and I think so still.

After fifteen months of this exercise, we went up together for our examination at Montpellier; and both of us received our degrees as bachelors of mathematical science. My companion was a wreck: I, on the other hand, had refreshed myself with analytical geometry.

Utterly worn out by his course of conic sections, my chum declares that he has had enough. In vain I hold out the glittering prospect of a new degree, that of licentiate of mathematical science, which would lead us to the splendors of the higher mathematics and initiate us into the mechanics of the heavens: I cannot prevail upon him, cannot make him share my audacity. He calls it a mad scheme, which will exhaust us and come to nothing. Without the advice of an experienced pilot, with no other compass than a book, which is not always very clear, because of its laconic adherence to set terms, our poor bark is bound to be wrecked on the first reef. One might as well put out to sea in a nutshell and defy the billows of the vasty deep. He does not use these actual words, but his gloomy estimate of the extreme difficulties to be encountered is enough to explain his refusal. I am quite free to go and break my neck in far countries; he is more prudent and will not follow me.

I suspect another reason, which the deserter does not confess. He has obtained the title needed for his plans. What does he care for the rest? Is it worth while to sit up late at night and wear one's self out in toil for the mere pleasure of learning? He must be a madman who, without the lure of profit, lends an ear to the blandishments of knowledge. Let us retreat into our shell, close our lid to the importunities of the light and lead the life of a mussel. There lies the secret of happiness. This philosophy is not mine. My curiosity sees in a stage accomplished no more than the preparation for a new stage towards the retreating unknown. My partner, therefore, leaves me. Henceforth, I am alone, alone and wretched. There is no one left with whom I can sit up and thresh the subject out in exhilarating discussion. There is no one near me to understand me, no one who can even passively oppose his ideas to mine and take part in the conflict whence the light will spring, even as a spark is born of the concussion of two flints. When a difficulty arises, steep as a cliff, there is no friendly shoulder to support me in my attempt to climb it. Alone, I have to cling to the roughness of the jagged rock, to fall, often, and pick myself up, covered with bruises, and renew the assault; alone, I must give my shout of triumph, without the least echo of encouragement, when, reaching the summit and broken in the effort, I am at last allowed to see a little way beyond.

My mathematical campaign will cost me much stubborn thought: I am aware of this after the first few lines of my book. I am entering upon the domain of the abstract, rough ground that can only be cleared by the insistent plow of reflection. The blackboard, excellent for the curves of analytical geometry studied in my friend's company, is now neglected. I prefer the exercise book, a quire of paper bound in a cover. With this confidant, which allows one to remain seated and rests the muscles of the legs, I can commune nightly under my lampshade, until a late hour, and keep going the forge of thought wherein the intractable problem is softened and hammered into shape.

My study table, the size of a pocket handkerchief, occupied on the right by the ink stand—a penny bottle—and on the left by the open exercise book, gives me just the room which I need to wield the pen. I love that little piece of furniture, one of the first acquisitions of my early married life. It is easily moved where you wish: in front of the window, when the sky is cloudy; into the discreet light of a corner, when the sun is troublesome. In winter, it allows you to come close to the hearth, where a log is blazing.

Poor little walnut board, I have been faithful to you for half a century and more. Ink-stained, cut and scarred with the penknife, you lend your support today to my prose as you once did to my equations. This variation in employment leaves you indifferent; your patient back extends the same welcome to the formulae of algebra and the formula of thought. I cannot boast this placidity; I find that the change has not increased my peace of mind; hunting for ideas troubles the brain even more than hunting for the roots of an equation.

You would never recognize me, little friend, if you could give a glance at my gray mane. Where is the cheerful face of former days, bright with enthusiasm and hope? I have aged, I have aged. And you, what a falling off, since you came to me from the dealer's, gleaming and polished and smelling so good with your beeswax! Like your master, you have wrinkles, often my work, I admit; for how many times, in my impatience, have I not dug my pen into you, when, after its dip in the muddy inkpot, the nib refused to write decently!

One of your corners is broken off; the boards are beginning to come loose. Inside you, I hear, from time to time, the plane of the death-watch, who despoils old furniture. From year to year, new galleries are excavated, endangering your solidity. The old ones show on the outside in the shape of tiny round holes. A stranger has seized upon the latter, excellent quarters, obtained without trouble. I see the impudent intruder run nimbly under my elbow and penetrate forthwith into the tunnel abandoned by the death-watch. She is after game, this slender huntress, clad in black, busy collecting wood lice for her grubs. A whole nation is devouring you, you old table; I am writing on a swarm of insects! No support could be more appropriate to my entomological notes.

What will become of you when your master is gone? Will you be knocked down for a franc, when the family come to apportion my poor spoils? Will you be turned into a stand for the pitcher beside the kitchen sink? Will you be the plank on which the cabbages are shredded? Or will my children, on the contrary, agree and say:

'Let us preserve the relic. It was where he toiled so hard to teach himself and make himself capable of teaching others; it was where he so long consumed his strength to find food for us when we were little. Let us keep the sacred plank.'

I dare not believe in such a future for you. You will pass into strange hands, O my old friend; you will become a bedside table, laden with bowl after bowl of linseed tea, until, decrepit, rickety and broken down, you are chopped up to feed the flames for a brief moment under the simmering saucepan. You will vanish in smoke to join my labors in that other smoke, oblivion, the ultimate resting place of our vain agitations.

But let us return, little table, to our young days; those of your shining varnish and of my fond illusions. It is Sunday, the day of rest, that is to say, of continuous work, uninterrupted by my duties in the school. I greatly prefer Thursday, which is not a general holiday and more propitious to studious calm. Such as it is, for all its distractions, the Lord's day gives me a certain leisure. Let us make the most of it. There are fifty-two Sundays in the year, making a total that is almost equivalent to the long vacation.

It so happens that I have a glorious question to wrestle with today; that of Kepler's three laws, which, when explored by the calculus, are to show me the fundamental mechanism of the heavenly bodies. One of them says: 'The area swept out in a given time by the radius vector of the path of a planet is proportional to the time taken.'

From this I have to deduce that the force which confines the planet to its orbit is directed towards the sun. Gently entreated by the differential and integral calculus, already the formula is beginning to voice itself. My concentration redoubles, my mind is set upon seizing the radiant dawn of truth.

Suddenly, in the distance, br-r-r-rum! Br-r-r-rum! Br-r-r-rum! The noise comes nearer, grows louder. Woe upon me! And plague take the Pagoda!

Let me explain. I live in a suburb, at the beginning of the Pernes Road, far from the tumult of the town [of Carpentras where Fabre was a master at the college]. Twenty yards in front of my house, some pleasure gardens have been opened, bearing a signboard inscribed, 'The Pagoda.' Here, on Sunday afternoons, the lads and lasses from the neighboring farms come to disport themselves in country dances. To attract custom and push the sale of refreshments, the proprietor of the ball ends the Sunday hop with a tombola. Two hours beforehand, he has the prizes carried along the public roads, preceded by fifes and drums. From a beribboned pole, borne by a stalwart fellow in a red sash, dangle a plated goblet, a handkerchief of Lyons silk, a pair of candlesticks and some packets of cigars. Who would not enter the pleasure gardens, with such a bait?

'Br-r-r-rum! Br-r-r-rum! Br-r-r-rum!' goes the procession.

It comes just under my window, wheels to the right and marches into the establishment, a huge wooden booth, hung with evergreens. And now, if you dislike noise, flee, flee as far as you can. Until nightfall, the ophicleides will bellow, the fifes tootle and the cornets bray. How would you deduce the steps of Kepler's laws to the accompaniment of that noisy orchestra! It is enough to drive one mad. Let us be off with all speed.

A mile away, I know a flinty waste beloved of the wheatear and the locust. Here reigns perfect calm; moreover, there are some clumps of evergreen oak which will lend me their scanty shade. I take my book, a few sheets of paper and a pencil and fly to this solitude. What beauteous silence, what exquisite quiet! But the sun is overwhelming, under the meager cover of the bushes. Cheerily, my lad! Have at your Kepler's laws in the company of the blue-winged locusts. You will return home with your problems solved, but with a blistered skin. An overdose of sun in the neck shall be the outcome of grasping the law of the areas. One thing makes up for another.

During the rest of the week, I have my Thursdays and the evenings, which I employ in study until I drop with sleep. All told I have no lack of time, despite the drudgery of my college ties. The great thing is not to be discouraged by the unavoidable difficulties encountered at the outset. I lose my way easily in that dense forest overgrown with creepers that have to be cut away with the axe to obtain a clearing. A fortunate turn or two; and I once more know where I am. I lose my way again. The stubborn axe makes its opening without always letting in sufficient light.

The book is just a book, that is to say, a set text, saying not a word more than it is obliged to, exceedingly learned, I admit, but, alas, often obscure! The author, it seems, wrote it for himself. He understood; therefore others must. Poor beginners, left to yourselves, you manage as best you can! For you, there shall be no retracing of steps in order to tackle the difficulty in another way; no circuit easing the arduous road and preparing the passage; no supplementary aperture to admit a glimmer of daylight. Incomparably inferior to the spoken word, which begins again with fresh methods of attack and is ready to vary the paths that lead to the open, the book says what it says and nothing more. Having finished its demonstration, whether you understand or no, the oracle is inexorably dumb. You reread the text and ponder it obstinately; you pass and repass your shuttle through the woof of figures. Useless efforts all: the darkness continues. What would be needed to supply the illuminating ray? Often enough, a trifle, a mere word; and that word the book will not speak.

Happy is he who is guided by a master's teaching! His progress does not know the misery of those wearisome breakdowns. What was I to do before the disheartening wall that every now and then rose up and barred my road? I followed d'Alembert's precept in his advice to young mathematical students: 'Have faith and go ahead,' said the great geometrician.

Faith I had; and I went on pluckily. And it was well for me that I did, for I often found behind the wall the enlightenment which I was seeking in front of it. Giving up the bad patch as hopeless, I would go on and, after I had left it behind, discover the dynamite capable of blasting it. 'Twas a tiny grain at first, an insignificant ball rolling and increasing as it went. From one slope to the other of the theorems, it grew to a heavy mass; and the mass became a mighty projectile which, flung backwards and retracing its course, split the darkness and spread it into one vast sheet of light.

D'Alembert's precept is good and very good, provided you do not abuse it. Too much precipitation in turning over the intractable page might expose you to many a disappointment. You must have fought the difficulty tooth and nail before abandoning it. This rough skirmishing leads to intellectual vigor.

Twelve months of meditation in the company of my little table at last won me my degree as a licentiate of mathematical science; and I was now qualified to perform, half a century later, the eminently lucrative functions of an inspector of Spiders' webs!

CHAPTER XIV.
THE BLUEBOTTLE: THE LAYING

To purge the earth of death's impurities and cause deceased animal matter to be once more numbered among the treasures of life there are hosts of sausage queens, including, in our part of the world, the bluebottle (Calliphora vomitaria, LIN.) and the checkered flesh fly (Sarcophaga carnaria, LIN.). Every one knows the first, the big, dark-blue fly who, after effecting her designs in the ill-watched meat safe, settles on our window panes and keeps up a solemn buzzing, anxious to be off in the sun and ripen a fresh emission of germs. How does she lay her eggs, the origin of the loathsome maggot that battens poisonously on our provisions, whether of game or butcher's meat? What are her stratagems and how can we foil them? This is what I propose to investigate.

The bluebottle frequents our homes during autumn and a part of winter, until the cold becomes severe; but her appearance in the fields dates back much earlier. On the first fine day in February, we shall see her warming herself, chillily, against the sunny walls. In April, I notice her in considerable numbers on the laurestinus. It is here that she seems to pair, while sipping the sugary exudations of the small white flowers. The whole of the summer season is spent out of doors, in brief flights from one refreshment bar to the next. When autumn comes, with its game, she makes her way into our houses and remains until the hard frosts.

This suits my stay-at-home habits and especially my legs, which are bending under the weight of years. I need not run after the subjects of my present study; they call on me. Besides, I have vigilant assistants. The household knows of my plans. Every one brings me, in a little screw of paper, the noisy visitor just captured against the panes.

Thus do I fill my vivarium, which consists of a large, bell-shaped cage of wire gauze, standing in an earthenware pan full of sand. A mug containing honey is the dining room of the establishment. Here the captives come to recruit themselves in their hours of leisure. To occupy their maternal cares, I employ small birds—chaffinches, linnets, sparrows—brought down, in the enclosure, by my son's gun.

I have just served up a Linnet shot two days ago. I next place in the cage a bluebottle, one only, to avoid confusion. Her fat belly proclaims the advent of a laying time. An hour later, when the excitement of being put in prison is allayed, my captive is in labor. With eager, jerky steps, she explores the morsel of game, goes from the head to the tail, returns from the tail to the head, repeats the action several times and at last settles near an eye, a dimmed eye sunk into its socket.

The ovipositor bends at a right angle and dives into the junction of the beak, straight down to the root. Then the eggs are emitted for nearly half an hour. The layer, utterly absorbed in her serious business, remains stationary and impassive and is easily observed through my lens. A movement on my part would doubtless scare her; but my restful presence gives her no anxiety. I am nothing to her.

The discharge does not go on continuously until the ovaries are exhausted; it is intermittent and performed in so many packets. Several times over, the fly leaves the bird's beak and comes to take a rest upon the wire gauze, where she brushes her hind legs one against the other. In particular, before using it again, she cleans, smoothes and polishes her laying tool, the probe that places the eggs. Then, feeling her womb still teeming, she returns to the same spot at the joint of the beak. The delivery is resumed, to cease presently and then begin anew. A couple of hours are thus spent in alternate standing near the eye and resting on the wire gauze.

At last, it is over. The fly does not go back to the bird, a proof that her ovaries are exhausted. The next day, she is dead. The eggs are dabbed in a continuous layer, at the entrance to the throat, at the root of the tongue, on the membrane of the palate. Their number appears considerable; the whole inside of the gullet is white with them. I fix a little wooden prop between the two mandibles of the beak, to keep them open and enable me to see what happens.

I learn in this way that the hatching takes place in a couple of days. As soon as they are born, the young vermin, a swarming mass, leave the place where they are and disappear down the throat. To inquire further into the work is useless for the moment. We shall learn more about it later, under conditions that make examination easier.

The beak of the bird invaded was closed at the start, as far as the natural contact of the mandibles allowed. There remained a narrow slit at the base, sufficient at most to admit the passage of a horsehair. It was through this that the laying was performed. Lengthening her ovipositor like a telescope, the mother inserted the point of her implement, a point slightly hardened with a horny armor. The fineness of the probe equals the fineness of the aperture. But, if the beak were entirely closed, where would the eggs be laid then?

With a tied thread, I keep the two mandibles in absolute contact; and I place a second bluebottle in the presence of the linnet, which the colonists have already entered by the beak. This time, the laying takes place on one of the eyes, between the lid and the eyeball. At the hatching, which again occurs a couple of days later, the grubs make their way into the fleshy depths of the socket. The eyes and the beak, therefore, form the two chief entrances into feathered game.

There are others; and these are the wounds. I cover the linnet's head with a paper hood which will prevent invasion through the beak and eyes. I serve it, under the wire gauze bell, to a third egg layer. The bird has been struck by a shot in the breast, but the sore is not bleeding: no outer stain marks the injured spot. Moreover, I am careful to arrange the feathers, to smooth them with a hair pencil, so that the bird looks quite smart and has every appearance of being untouched.

The fly is soon there. She inspects the linnet from end to end; with her front tarsi she fumbles at the breast and belly. It is a sort of auscultation by sense of touch. The insect becomes aware of what is under the feathers by the manner in which these react. If scent comes to her assistance, it can only be very slightly, for the game is not yet high. The wound is soon found. No drop of blood is near it, for it is closed by a plug of down rammed into it by the shot. The fly takes up her position without separating the feathers or uncovering the wound. She remains here for two hours without stirring, motionless, with her abdomen concealed beneath the plumage. My eager curiosity does not distract her from her business for a moment.

When she has finished, I take her place. There is nothing either on the skin or at the mouth of the wound. I have to withdraw the downy plug and dig to some depth before discovering the eggs. The ovipositor has therefore lengthened its extensible tube and pushed beyond the feather stopper driven in by the lead. The eggs are in one packet; they number about three hundred.

When the beak and eyes are rendered inaccessible, when the body, moreover, has no wounds, the laying still takes place, but, this time, in a hesitating and niggardly fashion. I pluck the bird completely, the better to watch what happens; also, I cover the head with a paper hood to close the usual means of access. For a long time, with jerky steps, the mother explores the body in every direction; she takes her stand by preference on the head, which she sounds by tapping on it with her front tarsi. She knows that the openings which she needs are there, under the paper; but she also knows how frail are her grubs, how powerless to pierce their way through the strange obstacle which stops her as well and interferes with the work of her ovipositor. The cowl inspires her with profound distrust. Despite the tempting bait of the veiled head, not an egg is laid on the wrapper, slight though it may be.

Weary of vain attempts to compass this obstacle, the Fly at last decides in favor of other points, but not on the breast, belly or back, where the hide would seem too tough and the light too intrusive. She needs dark hiding places, corners where the skin is very delicate. The spots chosen are the cavity of the axilla, corresponding with our armpit, and the crease where the thigh joins the belly. Eggs are laid in both places, but not many, showing that the groin and the axilla are adopted only reluctantly and for lack of a better spot.

With an unplucked bird, also hooded, the same experiment failed: the feathers prevent the fly from slipping into those deep places. Let us add, in conclusion, that, on a skinned bird, or simply on a piece of butcher's meat, the laying is effected on any part whatever, provided that it be dark. The gloomiest corners are the favorite ones.

It follows from all this that, to lay the eggs, the Bluebottle picks out either naked wounds or else the mucous membranes of the mouth or eyes, which are not protected by a skin of any thickness. She also needs darkness. We shall see the reasons for her preference later on.

The perfect efficiency of the paper bag, which prevents the inroads of the worms through the eye sockets or the beak, suggests a similar experiment with the whole bird. It is a matter of wrapping the body in a sort of artificial skin which will be as discouraging to the fly as the natural skin. Linnets, some with deep wounds, others almost intact, are placed one by one in paper envelopes similar to those in which the nursery gardener keeps his seeds, envelopes just folded, without being stuck. The paper is quite ordinary and of average thickness. Torn pieces of newspaper serve the purpose.

These sheaths with the corpses inside them are freely exposed to the air, on the table in my study, where they are visited, according to the time of day, in dense shade and in bright sunlight. Attracted by the effluvia from the dead meat, the bluebottles haunt my laboratory, the windows of which are always open. I see them daily alighting on the envelopes and very busily exploring them, apprised of the contents by the gamy smell. Their incessant coming and going is a sign of intense cupidity; and yet none of them decides to lay on the bags. They do not even attempt to slide their ovipositor through the slits of the folds. The favorable season passes and not an egg is laid on the tempting wrappers. All the mothers abstain, judging the slender obstacle of the paper to be more than the vermin will be able to overcome.

This caution on the fly's part does not at all surprise me: motherhood everywhere has gleams of great perspicacity. What does astonish me is the following result. The parcels containing the linnets are left for a whole year uncovered on the table; they remain there for a second year and a third. I inspect the contents from time to time. The little birds are intact, with unrumpled feathers, free from smell, dry and light, like mummies. They have become not decomposed, but mummified.

I expected to see them putrefying, running into sanies, like corpses left to rot in the open air. On the contrary, the birds have dried and hardened, without undergoing any change. What did they want for their putrefaction? Simply the intervention of the fly. The maggot, therefore, is the primary cause of dissolution after death; it is, above all, the putrefactive chemist.

A conclusion not devoid of value may be drawn from my paper game bags. In our markets, especially in those of the South, the game is hung unprotected from the hooks on the stalls. Larks strung up by the dozen with a wire through their nostrils, thrushes, plovers, teal, partridges, snipe, in short, all the glories of the spit which the autumn migration brings us, remain for days and weeks at the mercy of the flies. The buyer allows himself to be tempted by a goodly exterior; he makes his purchase and, back at home, just when the bird is being prepared for roasting, he discovers that the promised dainty is alive with worms. O horror! There is nothing for it but to throw the loathsome, verminous thing away.

The bluebottle is the culprit here. Everybody knows it; and nobody thinks of seriously shaking off her tyranny: not the retailer, nor the wholesale dealer, nor the killer of the game. What is wanted to keep the maggots out? Hardly anything: to slip each bird into a paper sheath. If this precaution were taken at the start, before the flies arrive, any game would be safe and could be left indefinitely to attain the degree of ripeness required by the epicure's palate.

Stuffed with olives and myrtle berries, the Corsican blackbirds are exquisite eating. We sometimes receive them at Orange, layers of them, packed in baskets through which the air circulates freely and each contained in a paper wrapper. They are in a state of perfect preservation, complying with the most exacting demands of the kitchen. I congratulate the nameless shipper who conceived the bright idea of clothing his blackbirds in paper. Will his example find imitators? I doubt it.

There is, of course, a serious objection to this method of preservation. In its paper shroud, the article is invisible; it is not enticing; it does not inform the passer by of its nature and qualities. There is one resource left which would leave the bird uncovered: simply to case the head in a paper cap. The head being the part most threatened, because of the mucus membrane of the throat and eyes, it would be sufficient, as a rule, to protect the head, in order to keep off the Flies and to thwart their attempts.

Let us continue to study the bluebottle, while varying our means of information. A tin, about four inches deep, contains a piece of butcher's meat. The lid is not put in quite straight and leaves a narrow slit at one point of its circumference, allowing, at most, of the passage of a fine needle. When the bait begins to give off a gamy scent, the mothers come. Singly or in numbers. They are attracted by the odor which, transmitted through a thin crevice, hardly reaches my nostrils.

They explore the metal receptacle for some time, seeking an entrance. Finding naught that enables them to reach the coveted morsel, they decide to lay their eggs on the tin, just beside the aperture. Sometimes, when the width of the passage allows of it, they insert the ovipositor into the tin and lay the eggs inside, on the very edges of the slit. Whether outside or in, the eggs are dabbed down in a fairly regular and absolutely white layer. I as it were shovel them up with a little paper scoop. I thus obtain all the germs that I require for my experiments, eggs bearing no trace of the stains which would be inevitable if I had to collect them on tainted meat.

We have seen the bluebottle refusing to lay her eggs on the paper bag, notwithstanding the carrion fumes of the Linnet enclosed; yet now, without hesitation, she lays them on a sheet of metal. Can the nature of the floor make any difference to her? I replace the tin lid by a paper cover stretched and pasted over the orifice. With the point of my knife, I make a narrow slit in this new lid. That is quite enough: the parent accepts the paper.

What determined her, therefore, is not simply the smell, which can easily be perceived even through the uncut paper, but, above all, the crevice, which will provide an entrance for the vermin, hatched outside, near the narrow passage. The maggots' mother has her own logic, her prudent foresight. She knows how feeble her wee grubs will be, how powerless to cut their way through an obstacle of any resistance; and so, despite the temptation of the smell, she refrains from laying so long as she finds no entrance through which the newborn worms can slip unaided.

I wanted to know whether the color, the shininess, the degree of hardness and other qualities of the obstacle would influence the decision of a mother obliged to lay her eggs under exceptional conditions. With this object in view, I employed small jars, each baited with a bit of butcher's meat. The respective lids were made of different colored paper, of oilskin, or of some of that tinfoil, with its gold or coppery sheen, which is used for sealing liqueur bottles. On not one of these covers did the mothers stop, with any desire to deposit their eggs; but, from the moment that the knife had made the narrow slit, all the lids were, sooner or later, visited and all of them, sooner or later, received the white shower somewhere near the gash. The look of the obstacle, therefore, does not count; dull or brilliant, drab or colored: these are details of no importance; the thing that matters is that there should be a passage to allow the grubs to enter.

Though hatched outside, at a distance from the coveted morsel, the newborn worms are well able to find their refectory. As they release themselves from the egg, without hesitation, so accurate is their scent, they slip beneath the edge of the ill-joined lid, or through the passage cut by the knife. Behold them entering upon their promised land, their reeking paradise.

Eager to arrive, do they drop from the top of the wall? Not they! Slowly creeping, they make their way down the side of the jar; they use their fore part, ever in quest of information, as a crutch and grapnel in one. They reach the meat and at once install themselves upon it.

Let us continue our investigation, varying the conditions. A large test-tube, measuring nine inches high, is baited at the bottom with a lump of butcher's meat. It is closed with wire gauze, whose meshes, two millimeters wide, do not permit of the fly's passage. The bluebottle comes to my apparatus, guided by scent rather than sight. She hastens to the test tube whose contents are veiled under an opaque cover with the same alacrity as to the open tube. The invisible attracts her quite as much as the visible.

She stays a while on the lattice of the mouth, inspects it attentively; but, whether because circumstances have failed to serve me, or because the wire network inspires her with distrust, I never saw her dab her eggs upon it for certain. As her evidence was doubtful, I had recourse to the flesh fly (Sarcophaga carnaria).

This fly is less finicky in her preparations, she has more faith in the strength of her worms, which are born ready-formed and vigorous, and easily shows me what I wish to see. She explores the trellis-work, chooses a mesh through which she inserts the tip of her abdomen and, undisturbed by my presence, emits, one after the other, a certain number of grubs, about ten or so. True, her visits will be repeated, increasing the family at a rate of which I am ignorant.

The newborn worms, thanks to a slight viscidity, cling for a moment to the wire gauze; they swarm, wriggle, release themselves and leap into the chasm. It is a nine inch drop at least. When this is done, the mother makes off, knowing for a certainty that her offspring will shift for themselves. If they fall on the meat, well and good; if they fall elsewhere, they can reach the morsel by crawling.

This confidence in the unknown factor of the precipice, with no indication but that of smell, deserves fuller, investigation. From what height will the flesh fly dare to let her children drop? I top the test-tube with another tube, the width of the neck of a claret bottle. The mouth is closed either with wire gauze, or with a paper cover with a slight cut in it. Altogether, the apparatus measures twenty-five inches in height. No matter: the fall is not serious for the lithe backs of the young grubs; and, in a few days, the test-tube is filled with larvae, in which it is easy to recognize the flesh fly's family by the fringed coronet that opens and shuts at the maggot's stern like the petals of a little flower. I did not see the mother operating: I was not there at the time; but there is no doubt possible of her coming nor of the great dive taken by the family: the contents of the test-tube furnish me with a duly authenticated certificate.

I admire the leap and, to obtain one better still, I replace the tube by another, so that the apparatus now stands forty-six inches high. The column is erected at a spot frequented by flies, in a dim light. Its mouth, closed with a wire gauze cover, reaches the level of various other appliances, test-tubes and jars, which are already stocked or awaiting their colony of vermin. When the position is well known to the flies, I remove the other tubes and leave the column, lest the visitors should turn aside to easier ground.

From time to time, the bluebottle and the flesh fly perch on the trellis-work, make a short investigation and then decamp. Throughout the summer season, for three whole months, the apparatus remains where it is, without the least result: never a worm. What is the reason? Does the stench of the meat not spread, coming from that depth? Certainly it spreads: it is unmistakable to my dulled nostrils and still more so to the nostrils of my children, whom I call to bear witness. Then why does the flesh fly, who but now was dropping her grubs from a goodly height, refuse to let them fall from the top of a column twice as high? Does she fear lest her worms should be bruised by an excessive drop? There is nothing about her to point to anxiety aroused by the length of the shaft. I never see her explore the tube or take its size. She stands on the trellised orifice; and there the matter ends. Can she be apprised of the depth of the chasm by the comparative faintness of the offensive odors that arise from it? Can the sense of smell measure the distance and judge whether it be acceptable or not? Perhaps.

The fact remains that, despite the attraction of the scent, the flesh fly does not expose her worms to disproportionate falls. Can she know beforehand that, when the chrysalides break, her winged family, knocking with a sudden flight against the sides of a tall chimney, will be unable to get out? This foresight would be in agreement with the rules which order maternal instinct according to future needs.

But when the fall does not exceed a certain depth, the budding worms of the flesh fly are dropped without a qualm, as all our experiments show. This principle has a practical application which is not without its value in matters of domestic economy. It is as well that the wonders of entomology should sometimes give us a hint of commonplace utility.

The usual meat safe is a sort of large cage with a top and bottom of wood and four wire gauze sides. Hooks fixed into the top are used whereby to hang pieces which we wish to protect from the flies. Often, so as to employ the space to the best advantage, these pieces are simply laid on the floor on the cage. With these arrangements, are we sure of warding off the fly and her vermin?

Not at all. We may protect ourselves against the Bluebottle, who is not much inclined to lay her eggs at a distance from the meat; but there is still the flesh fly, who is more venturesome and goes more briskly to work and who will slip the grubs through a hole in the meshes and drop them inside the safe. Agile as they are and well able to crawl, the worms will easily reach anything on the floor; the only things secure from their attacks will be the pieces hanging from the ceiling. It is not in the nature of maggots to explore the heights, especially if this implies climbing down a string in addition.

People also use wire gauze dish covers. The trellised dome protects the contents even less than does the meat safe. The flesh fly takes no heed of it. She can drop her worms through the meshes on the covered joint.

Then what are we to do? Nothing could be simpler. We need only wrap the birds which we wish to preserve—thrushes, partridges, snipe and so on—in separate paper envelopes; and the same with our beef and mutton. This defensive armor alone, while leaving ample room for the air to circulate, makes any invasion by the worms impossible, even without a cover or a meat safe: not that paper possesses any special preservative virtues, but solely because it forms an impenetrable barrier. The Bluebottle carefully refrains from laying her eggs upon it and the flesh fly from bringing forth her offspring, both of them knowing that their newborn young are incapable of piercing the obstacle.

Paper is equally successful in our strife against the Moths, those plagues of our furs and clothes. To keep away these wholesale ravages, people generally use camphor, naphthalene, tobacco, bunches of lavender and other strong-scented remedies. Without wishing to malign those preservatives, we are bound to admit that the means employed are none too effective. The smell does very little to prevent the havoc of the moths.

I would therefore counsel our housewives, instead of all this chemist's stuff, to use newspapers of a suitable shape and size. Take whatever you wish to protect—your furs, your flannel or your clothes—and pack each article carefully in a newspaper, joining the edges with a double fold, well pinned. If this joining is properly done, the Moth will never get inside. Since my advice has been taken and this method employed in my household, the old damage has never been repeated.

To return to the fly. A piece of meat is hidden in a jar under a layer of fine, dry sand, a finger's-breadth thick. The jar has a wide mouth and is left quite open. Let whoever come that will, attracted by the smell. The Bluebottles are not long in inspecting what I have prepared for them: they enter the jar, go out and come back again, inquiring into the invisible thing revealed by its fragrance. A diligent watch enables me to see them fussing about, exploring the sandy expanse, tapping it with their feet, sounding it with their proboscis. I leave the visitors undisturbed for a fortnight or three weeks. None of them lays any eggs.

This is a repetition of what the paper bag, with its dead bird, showed me. The flies refuse to lay on the sand, apparently for the same reasons. The paper was considered an obstacle which the frail vermin would not be able to overcome. With sand, the case is worse. Its grittiness would hurt the newborn weaklings, its dryness would absorb the moisture indispensable to their movements. Later, when preparing for the metamorphosis, when their strength has come to them, the grubs will dig the earth quite well and be able to descend; but, at the start, that would be very dangerous for them. Knowing these difficulties, the mothers, however greatly tempted by the smell, abstain from breeding. As a matter of fact, after long waiting, fearing lest some packets of eggs may have escaped my attention, I inspect the contents of the jar from top to bottom. Meat and sand contain neither larvae nor pupae: the whole is absolutely deserted.

The layer of sand being only a finger's-breadth thick, this experiment requires certain precautions. The meat may expand a little, in going bad, and protrude in one or two places. However small the fleshy eyots that show above the surface, the flies come to them and breed. Sometimes also the juices oozing from the putrid meat soak a small extent of the sandy floor. That is enough for the maggot's first establishment. These causes of failure are avoided with a layer of sand about an inch thick. Then the bluebottle, the flesh fly and other flies whose grubs batten on dead bodies are kept at a proper distance.

In the hope of awakening us to a proper sense of our insignificance, pulpit orators sometimes make an unfair use of the grave and its worms. Let us put no faith in their doleful rhetoric. The chemistry of man's final dissolution is eloquent enough of our emptiness: there is no need to add imaginary horrors. The worm of the sepulchre is an invention of cantankerous minds, incapable of seeing things as they are. Covered by but a few inches of earth, the dead can sleep their quiet sleep: no fly will ever come to take advantage of them.

At the surface of the soil, exposed to the air, the hideous invasion is possible; ay, it is the invariable rule. For the melting down and remolding of matter, man is no better, corpse for corpse, than the lowest of the brutes. Then the fly exercises her rights and deals with us as she does with any ordinary animal refuse. Nature treats us with magnificent indifference in her great regenerating factory: placed in her crucibles, animals and men, beggars and kings are one and all alike. There you have true equality, the only equality in this world of ours: equality in the presence of the maggot.

CHAPTER XV. THE BLUEBOTTLE: THE GRUB

The larvae of the bluebottle hatch within two days in the warm weather. Whether inside my apparatus, in direct contact with the piece of meat, or outside, on the edge of a slit that enables them to enter, they set to work at once. They do not eat, in the strict sense of the word, that is to say, they do not tear their food, do not chew it by means of implements of mastication. Their mouth parts do not lend themselves to this sort of work. These mouth parts are two horny spikes, sliding one upon the other, with curved ends that do not face, thus excluding the possibility of any function such as seizing and grinding.

The two guttural grapnels serve for walking much rather than for feeding. The worm plants them alternately in the road traversed and, by contracting its crupper, advances just that distance. It carries in its tubular throat the equivalent of our iron tipped sticks which give support and assist progress.

Thanks to this machinery of the mouth, the maggot not only moves over the surface, but also easily penetrates the meat: I see it disappear as though it were dipping into butter. It cuts its way, levying, as it goes, a preliminary toll, but only of liquid mouthfuls. Not the smallest solid particle is detached and swallowed. That is not the maggot's diet. It wants a broth, a soup, a sort of fluid extract of beef which it prepares itself. As digestion, after all, merely means liquefaction, we may say, without being guilty of paradox, that the grub of the bluebottle digests its food before swallowing it.

With the object of relieving gastric troubles, our manufacturing chemists scrape the stomachs of the pig and sheep and thus obtain pepsin, a digestive agent which possesses the property of liquefying albuminous matters and lean meat in particular. Why cannot they rasp the stomach of the maggot! They would obtain a product of the highest quality, for the carnivorous worm also owns its pepsin, pepsin of a singularly active kind, as the following experiments will show us.

I divide the white of a hard-boiled egg into tiny cubes and place them in a little test-tube. On the top of the contents, I sprinkle the eggs of the bluebottle, eggs free from the least stain, taken from those laid on the outside of tins baited with meat and not absolutely shut. A similar test-tube is filled with white of egg, but receives no germs. Both are closed with a plug of cotton-wool and left in a dark corner.

In a few days, the tube swarming with newborn vermin contains a liquid as fluid and transparent as water. Not a drop would remain in the tube if I turned it upside down. All the white of egg has disappeared, liquefied. As for the worms, which are already a fair size, they seem very ill at ease. Deprived of a support whence to attain the outer air, most of them dive into the broth of their own making, where they perish by drowning. Others, endowed with greater vigor, crawl up the glass to the plug and manage to make their way through the wadding. Their pointed front, armed with grappling irons, is the nail that penetrates the fibrous mass.

In the other test-tube, standing beside the first and subjected to the same atmospheric influences, nothing striking has occurred. The hard-boiled white of egg has retained its dead white color and its firmness. I find it as I left it. The utmost that I observe is a few traces of must. The result of this first experiment is patent: the Bluebottle's grub is the medium that converts coagulated albumen into a liquid.

The value of chemist's pepsin is estimated by the quantity of hard-boiled white of egg which a gram of that agent can liquefy. The mixture has to be exposed in an oven to a temperature of 1400 F. and also to be frequently shaken. My preparation, in which the bluebottle's eggs are hatched, is neither shaken nor subjected to the heat of an oven; everything happens in quietness and under the thermometric conditions of the surrounding air; nevertheless, in a few days, the coagulated albumen, treated by the vermin, runs like water.

The reagent that causes this liquefaction escapes my endeavors to detect it. The worms must disgorge it in infinitesimal doses, while the spikes in their throats, which are in continual movement, emerge a little way from the mouth, reenter and reappear. Those piston thrusts, those quasi-kisses, are accompanied by the emission of the solvent: at least, that is how I picture it. The maggot spits on its food, places on it the wherewithal to make it into broth. To appraise the quantity of the matter expectorated is beyond my powers: I observe the result, but do not perceive the leavening agent.

Well, this result is really astounding, when we consider the scantiness of the means. No pig's or sheep's pepsin can rival that of the worm. I have a bottle of pepsin that comes from the School of Chemistry at Montpellier. I lavishly powder some pieces of hard-boiled white of egg with the potent drug, just as I did with the eggs of the Bluebottle. The oven is not brought into play, neither is distilled water added, nor hydrochloric acid: two auxiliaries which are recommended. The experiment is conducted in exactly the same way as that of the tubes with the vermin. The result is entirely different from what I expected. The white of egg does not liquefy. It simply becomes moist on the surface; and even this moisture may come from the pepsin, which is highly absorbent. Yes, I was right: if the thing were feasible, it would be an advantage for the chemists to collect their digestive drug from the stomach of the maggot. The worm, in this case, beats the pig and the sheep.

The same method is followed for the remaining experiments. I put the bluebottle's eggs to hatch on a piece of meat and leave the worms to do their work as they please. The lean tissues, whether of mutton, beef or pork, no matter which, are not turned into liquid; they become a pea soup of a clarety brown. The liver, the lung, the spleen are attacked to better purpose, without, however, getting beyond the state of a semi-fluid jam, which easily mixes with water and even appears to dissolve in it. The brains do not liquefy either: they simply melt into a thin gruel.

On the other hand, fatty substances, such as beef suet, lard and butter, do not undergo any appreciable change. Moreover, the worms soon dwindle away, incapable of growing. This sort of food does not suit them. Why? Apparently because it cannot be liquefied by the reagent disgorged by the worms. In the same way, ordinary pepsin does not attack fatty substances; it takes pancreatin to reduce them to an emulsion. This curious analogy of properties, positive for albuminous, negative for fatty matter, proclaims the similarity and perhaps the identity of the dissolvent discharged by the grubs and the pepsin of the higher animals.

Here is another proof: the usual pepsin does not dissolve the epidermis, which is a material of a horny nature. That of the maggots does not dissolve it either. I can easily rear bluebottle grubs on dead crickets whose bellies I have first opened; but I do not succeed if the morsel be left intact: the worms are unable to perforate the succulent paunch; they are stopped by the cuticle, on which their reagent refuses to act. Or else I give them frogs' hind legs, stripped of their skin. The flesh turns to broth and disappears to the bone. If I do not peel the legs, they remain intact in the midst of the vermin. Their thin skin is sufficient to protect them.

This failure to act upon the epidermis explains why the bluebottle at work on the animal declines to lay her eggs on the first part that comes handy. She needs the delicate membrane of the nostrils, eyes or throat, or else some wound in which the flesh is laid bare. No other place suits her, however excellent for flavor and darkness. At most, finding nothing better when my stratagems interfere, she persuades herself to dab a few eggs under the axilla of a plucked bird or in the groin, two points at which the skin is thinner than elsewhere.

With her maternal foresight, the bluebottle knows to perfection the choice surfaces, the only ones liable to soften and run under the influence of the reagent dribbled by the newborn grubs. The chemistry of the future is familiar to her, though she does not use it for her own feeding; motherhood, that great inspirer of instinct, teaches her all about it.

Scrupulous though she be in choosing exactly where to lay her eggs, the bluebottle does not trouble about the quality of the provisions intended for her family's consumption. Any dead body suits her purpose. Redi, the Italian scientist who first exploded the old, foolish notion of worms begotten of corruption, fed the vermin in his laboratory with meat of very different kinds. In order to make his tests the more conclusive, he exaggerated the largess of the dining hall. The diet was varied with tiger and lion flesh, bear and leopard, fox and wolf, mutton and beef, horseflesh, donkey flesh and many others, supplied by the rich menagerie of Florence. This wastefulness was unnecessary: wolf and mutton are all the same to an unprejudiced stomach.

A distant disciple of the maggot's biographer, I look at the problem in a light which Redi never dreamt of. Any flesh of one of the higher animals suits the fly's family. Will it be the same if the food supplied be of a lower organism and consist of fish, for instance, of frog, mollusk, insect, centipede? Will the worms accept these viands and, above all, can they manage to liquefy them, which is the first and foremost condition?

I serve a piece of raw whiting. The flesh is white, delicate, partly translucent, easy for our stomachs to digest and no less suited to the grub's dissolvent. It turns into an opalescent fluid, which runs like water. In fact, it liquefies in much the same way as hard-boiled white of egg. The worms at first wax fat, as long as the conditions allow of some solid eyots remaining; then, when foothold fails, threatened with drowning in the too fluid broth, they creep up the side of the glass, anxious and restless to be off. They climb to the cotton-wool stopper of the test-tube and try to bolt through the wadding. Endowed with stubborn perseverance, nearly all of them decamp in spite of the obstacle. The test-tube with the white of egg showed me a similar exodus. Although the fare suits them, as their growth witnesses, the worms cease feeding and make a point of escaping when death by drowning is imminent.

With other fish, such as skate and sardines, with the flesh of frogs and tree frogs, the meat simply dissolves into a porridge. Hashes of slug, Scolopendra or praying mantis furnish the same result.

In all these preparations, the dissolving agent of the worms is as much in evidence as when butcher's meat is employed. Moreover, the grubs seem satisfied with the queer dish which my curiosity prescribes for them; they thrive amidst the victuals and undergo their transformation into pupae.

The conclusion, therefore, is much more general than Redi imagined. Any meat, no matter whether of a higher or lower order, suits the bluebottle for the settlement of her family. The carcasses of furred and feathered animals are the favorite victuals, probably because of their richness, which allows of plentiful layings; but, should the occasion demand it, the others are also accepted, without inconvenience. Any carrion that has lived the life of an animal comes within the domain of these scavengers.

What is their number to one mother? I have already spoken of a deposit of three hundred, counted egg by egg. A quite fortuitous circumstance enabled me to go much farther. In the first week of January 1905, we experienced a sudden short cold snap of a severity very exceptional in my part of the country. The thermometer fell to twelve degrees below zero. While a fierce north wind was raging and beginning to redden the leaves of the olive trees, came one and brought me a barn or screech owl, which he had found on the ground, exposed to the air, not far from my house. My reputation as a lover of animals made the donor believe that I should be pleased with his gift.

I was, as a matter of fact, but for reasons whereof the finder certainly never dreamt. The owl was untouched, with trim feathers and not the least wound that showed. Perhaps he had died of cold. What made me gratefully accept the present was exactly that which would have inclined anyone but myself to refuse it. The owl's eyes, glazed in death, were hidden under a thick mass of eggs, which I recognized as a bluebottle's. Similar masses occupied the vicinity of the nostrils. If I wanted maggots, here, of a certainty, was a richer crop than I had ever beheld.

I place the corpse on the sand of a pan, with a wire gauze cover, and leave events to take their course. The laboratory in which I install my bird is none other than my study. It is as cold in there, or nearly, as outside, so much so that the water in the aquarium in which I used to rear caddis worms has frozen into a solid block of ice. Under these conditions of temperature, the owl's eyes keep their white veil of germs unchanged. Nothing stirs, nothing swarms. Weary of waiting, I pay no more attention to the carcass; I leave the future to decide whether the cold has exterminated the fly's family or not.

Before the end of March, the packets of eggs have disappeared, I know not how long. The bird, for that matter, seems to be intact. On the ventral surface, which is turned to the air, the feathers keep their smooth arrangement and their fresh coloring. I lift the thing. It is light, very dry and gives a hard sound, like an old shoe tanned by the summer sun in the fields. There is no smell. The dryness has vanquished the stench, which, in any case, was never offensive during that time of frost. On the other hand, the back, which touched the sand, is a loathsome wreck, partly deprived of its feathers. The quills of the tail are bare barreled; a few whitened bones show, deprived of their muscles. The skin has turned into a dark leather, pierced with round holes like those of a sieve. It is all hideously ugly, but most instructive.

The wretched owl, with his shattered backbone, teaches us, first of all, that a temperature twelve degrees of frost does not endanger the existence of the bluebottle's germs. The worms were born without accident, despite the rude blast; they feasted copiously on extract of meat; then, growing big and fat, they descended into the earth by piercing round holes in the bird's skin. Their pupae must now be in the sand of the pan.

They are, in point of fact, and in such numbers that I have to resort to sifting in order to collect them. If I used the forceps, I should never have done sorting so great a quantity. The sand passes through the meshes of the sieve, the pupae remain above. To count them would wear out my patience. I measure them by the bushel, that is to say, with a thimble of which I know the holding capacity in pupae. The result of my calculation is not far short of nine hundred.

Does this family proceed from one mother? I am quite ready to admit it, so unlikely is it that the bluebottle, who is so rare inside our houses during the severe cold of winter, should be frequent enough outside to form into groups and to do business in common while an icy blast is raging. A belated specimen, the plaything of the north wind, and one alone must have deposited the burden of her ovaries on the owl's eyes. This laying of nine hundred eggs, an incomplete laying perhaps, bears witness to the mighty part played by the fly as a liquidator of corpses.

Before throwing away the screech owl treated by the worms, let us overcome our repugnance and give a glance inside the bird. We see a tortuous cavity, fenced in by nameless ruins. Muscles and bowels have disappeared, converted into broth and gradually consumed by the teeming throng. In every part, what was wet has become dry, what was solid muddy. In vain my forceps ransacks every nook and corner: it does not hit upon a single pupa. All the worms have emigrated, all, without exception. From first to last, they have forsaken the refuge of the corpse, so soft to their delicate skins; they have left the velvet for the hard ground. Is dryness necessary to them at this stage? They had it in the carcass, which was thoroughly drained. Would they protect themselves against the cold and rain? No shelter could suit them better than the thick quilt of the feathers, which has remained wholly undamaged on the belly, the breast and every part that was not in touch with the ground. It looks as though they had fled from comfort to seek a less kindly dwelling place. When the hour of transformation came, all left the owl, that most excellent lodging; all dived into the sand.

The exodus from the mortuary tabernacle was made through the round holes wherewith the skin is pierced. Those holes are the worms' work: of that there is no doubt; and yet we have lately seen the mothers refuse as a bed for their eggs any part whereat the flesh is protected by a skin of some thickness. The reason is the failure of the pepsin to act on epidermic substances. In the absence of liquefaction at such points, the nourishing gruel is unprocurable. On the other hand, the tiny worms are not able—or at least do not know how—to dig through the integument with their pair of guttural harpoons, to rend it and reach the liquefiable flesh. The newborn lack strength and, above all, purpose. But, as the time comes for descending into the earth, the worms, now powerful and suddenly versed in the necessary art, well know how to eat away patiently and clear themselves a passage. With the hooks of their spikes they dig, scratch and tear. Instinct has flashes of inspiration. What the animal did not know how to do at the start it learns without apprenticeship when the time comes to practice this or that industry. The maggot ripe for burial perforates a membranous obstacle which the grub intent upon its broth would not even have attempted to attack with either its pepsin or its grapnels.

Why does the worm quit the carcass, that capital shelter? Why does it go and take up its abode in the ground? As the leading disinfector of dead things, it works at the most important matter, the suppression of the infection; but it leaves a plentiful residuum, which does not yield to the reagents of its analytical chemistry. These remains have to disappear in their turn. After the fly, anatomists come hastening, who take up the dry relic, nibble skin, tendons and ligaments and scrape the bones clean.

The greatest expert in this work is the Dermestes beetle, an enthusiastic gnawer of animal remains. Sooner or later, he will come to the joint already exploited by the fly. Now what would happen if the pupae were there? The answer is obvious. The Dermestes, who loves hard food, would dig his teeth into the horny little kegs and demolish them at a bite. Even though he did not touch the contents, a live thing which he probably dislikes, he would at least test the flavor of that lifeless substance, the container. The future Fly would be lost, because her casing would be pierced. Even so, in the storerooms of our silk mills, a certain Dermestes (Dermestes vulpinus, FABR.) digs into the cocoons to attack the horny covering of the chrysalis.

The maggot foresees the danger and makes itself scarce before the other arrives. In what sort of memory does it house so much wisdom, indigent, headless creature that it is, for it is only by extension that we can give the name of head to the animal's pointed fore part? How did it learn that, to safeguard the pupa, it must desert the carcass and that, to safeguard the fly, it must not bury itself too far down?

To emerge from underground after the perfect insect is hatched, the bluebottle's device consists in disjointing her head into two movable halves, which, each distended with its great red eye, by turns separate and reunite. In the intervening space, a large, glassy hernia rises and disappears, disappears and rises. When the two move asunder, with one eye forced back to the right, the other to the left, it is as though the insect were splitting its brain pan in order to expel the contents. Then the hernia rises, blunt at the end and swollen into a great knob. Next, the forehead closes and the hernia retreats, leaving visible only a kind of shapeless muzzle. In short, a frontal pouch, with deep pulsations momentarily renewed, becomes the instrument of deliverance, the pestle wherewith the newly hatched bluebottle bruises the sand and causes it to crumble. Gradually the legs push the rubbish back and the insect advances so much toward the surface.

A hard task, this exhumation by dint of the blows of a cleft and palpitating head. Moreover, the exhausting effort has to be made at the moment of greatest weakness, when the insect leaves that protecting casket, its pupa. It emerges from it pale, flabby and unsightly, sorrily clad in the wings which, folded lengthwise and made shorter by their scalloped edge, only just cover the top of the back. Wildly bristling with hairs and colored ashen-gray, it is a piteous sight. The large set of wings, suitable for flight, will spread later. For the moment, it would only be in the way amid the obstacles to be passed through. Later also will come the faultless dress wherein the iridescent indigo-blue stands out against the severity of the black.

The frontal hernia that crumbles the sand with its impact has a tendency to make play for some time after the emergence from the ground. Take hold with the forceps of one of the hind legs of a newly released fly. Forthwith, the implement of the head begins to work, swelling and subsiding as energetically as a moment ago, when it had to make a hole in the sand. The insect, hampered in its movements as when it was underground, struggles as best it can against the only obstacle that it knows. With its heaving knob, it pounds the air even as but now it pounded the earthy barrier. In all unpleasant circumstances, its one resource is to cleave its head and produce its cranial hernia, which moves out and in, in and out. For nearly two hours, interspersed with halts due to fatigue, the little machine keeps throbbing in my forceps.

In the meantime, however, the desperate one is hardening her skin; she spreads wide the sail of her wings and dons her deep mourning of black and darkest blue. Then her eyes, warped sideways, come together and resume their normal position. The cleft forehead closes; the delivering blister goes in, never to show itself again. But there is one precaution to be taken first. With its front tarsi, the insect carefully brushes the bump about to disappear from view, lest grit should lodge in the cranium when the two halves of the head are joined for good.

The maggot is aware of the trials that await it when, as a fly, it will have to come up from under ground; it knows beforehand how difficult the ascent will be with the feeble instrument at its disposal, so difficult, in fact, as to become fatal should the journey be at all prolonged. It foresees the dangers ahead of it and averts them as well as it can. Gifted with two iron shod sticks in its throat, it can easily descend to such depths as it pleases. The need for greater quiet and a less trying temperature calls for the deepest possible home: the lower down it is, the better for the welfare of the worm and the pupa, on condition that descent be practicable. It is, perfectly; and yet, though free to obey its inspiration, the grub refrains. I rear it in a deep pan, full of fine, dry sand, easy to excavate. The interment never goes very far. About a hand's breadth is all that the most progressive digger ventures upon. Most of the interred remain nearer still to the surface. Here, under a thin layer of sand, the grub's skin hardens and becomes a coffin, a casket, wherein the transformation sleep is slept. A few weeks later, the buried one awakes, transfigured but weak, having naught wherewith to unearth herself but the throbbing hernia of her open forehead.

What the maggot denies itself it is open to me to realize, should I care to know the depth whence the fly is able to mount. I place fifteen bluebottle pupae, obtained in winter, at the bottom of a wide tube closed at one end. Above the pupae is a perpendicular column of fine, dry sand, the height of which varies in different tubes. April comes and the hatching begins.

A tube with six centimeters of sand, the shallowest of the columns under experiment, yields the best result. Of the fifteen subjects interred in the pupa stage, fourteen easily reach the surface when they become flies. Only one of them perishes, one who has not even attempted the ascent. With twelve centimeters of sand, four emerge. With twenty centimeters, two, no more. The other flies, jaded with their exertions, have died at a higher or lower stage of the road. Lastly, with yet another tube wherein the column of sand measured sixty centimeters, I obtained the liberation of only a single fly. The plucky creature must have had a hard struggle to mount from so great a depth, for the other fourteen did not even manage to burst the lid of their caskets.

I presume that the looseness of the sand and the consequent pressure in every direction, similar to that exercised by fluids, have a certain bearing on the difficulties of the exhumation. Two more tubes are prepared, but this time supplied with fresh mould, lightly heaped up, which has not the incoherence of sand, with the attendant drawback of pressure. Six centimeters of mould give me eight flies for fifteen pupae buried; twenty centimeters give me only one. There is less success than with the sandy column. My device has diminished the pressure, but, at the same time, increased the passive resistance. The sand falls of itself under the impact of the frontal rammer; the unyielding mould demands the cutting of a gallery. In fact, I perceive, on the road followed, a shaft which continues indefinitely such as it is. The fly has bored it with the temporary blister that throbs between her eyes.

In every medium, therefore, whether sand, mould or any earthy combination, great are the sufferings that attend the exhumation of the fly. And so the maggot shuns the depths which a desire for additional security might seem to recommend. The worm has its own prudence: foreseeing the dangers ahead, it refrains from making great descents that might promote the welfare of the moment. It neglects the present for the sake of the future.

CHAPTER XVI. A PARASITE OF THE MAGGOT

The dangers of the exhumation are not the only ones; the Bluebottle must be acquainted with others. Life, when all is said, is a knacker's yard wherein the devourer of today becomes the devoured of tomorrow; and the robber of the dead cannot fail to be robbed of her own life when the time comes. I know that she has one exterminator in the person of the tiny Saprinus beetle, a fisher of fat sausages on the edge of the pools formed by liquescent corpses. Here swarm in common the grubs of the greenbottle, the flesh fly and the bluebottle. The Saprinus draws them to him from the bank and gobbles them indiscriminately. They represent to him morsels of equal value.

This banquet can be observed only in the open country, under the rays of a hot sun. Saprini and greenbottles never enter our houses; the flesh fly visits us but discreetly, does not feel at home with us; the only one who comes fussing along is the bluebottle, who thus escapes the tribute due to the consumer of plump sausages. But, in the fields, where she readily lays her eggs upon any carcass that she finds, she, as well as the others, sees her vermin swept away by the gluttonous Saprinus.

In addition, graver disasters decimate her family, if, as I do not doubt, we can apply to the bluebottle what I have seen happen in the case of her rival, the flesh fly. So far, I have had no opportunity of actually perceiving with the first what I have to tell of the second; still, I do not hesitate to repeat about the one what observation has taught me about the other, for the larval analogies between the two flies are very close.

Here are the facts. I have gathered a number of pupae of the flesh fly in one of my vermin jars. Wishing to examine the pupa's hinder end, which is hollowed into a cup and scalloped into a coronet, I stave in one of the little barrels and force open the last segments with the point of my pocketknife. The horny keg does not contain what I expected to find: it is full of tiny grubs packed one atop the other with the same economy of space as anchovies in a bottle. Save for the skin, which has hardened into a brown shell, the substance of the maggot has disappeared, changed into a restless swarm.

There are thirty-five occupants. I replace them in their casket. The rest of my harvest, wherein, no doubt, are other pupae similarly stocked, is arranged in tubes that will easily show me what happens. The thing to discover is what genus of parasites the grubs enclosed belong to. But it is not difficult, without waiting for the hatching of the adults, to recognize their nature merely by their mode of life. They form part of the family of Chalcididae, who are microscopic ravagers of living entrails.

Not long ago, in winter, I took from the chrysalis of a great peacock moth four hundred and forty-nine parasites belonging to the same group. The whole substance of the future moth had disappeared, all but the nymphal wrapper, which was intact and formed a handsome Russia-leather wallet. The worm grubs were here heaped up and squeezed together to the point of sticking to one another. The hair pencil extracts them in bundles and cannot separate them without some difficulty. The holding capacity is strained to the utmost; the substance of the vanished Moth would not fill it better. That which died has been replaced by a living mass of equal dimensions, but subdivided. The price of this colony's existence is the conversion of the chrysalis into a sort of milk food of doubtful constitution. The enormous udder has been drained outright.

You shudder when you think of that budding flesh nibbled bit by bit by four or five hundred gormandizers; the horrified imagination refuses to picture the anguish suffered by the tortured wretch. But is there really any pain? We have leave to doubt it. Pain is a patent of nobility; it is more pronounced in proportion as the sufferer belongs to a higher order. In the lower ranks of animal life, it must be greatly reduced, perhaps even nil, especially when life, in the throes of evolution, has not yet acquired a stable equilibrium. The white of an egg is living matter, but endures the prick of a needle without a quiver. Would it not be the same with the chrysalis of the great peacock, dissected cell by cell by hundreds of infinitesimal anatomists? Would it not be the same with the pupa of the flesh fly? These are organisms put back into the crucible, reverting to the egg state for a second birth. There is reason to believe, therefore, that their destruction crumb by crumb is merciful.

Towards the end of August, the parasite of the flesh fly's grubs makes her appearance out of doors in the adult form. She is a Chalcidid, as I expected. She issues from the barrel through one or two little round holes which the prisoners have pierced with a patient tooth. I count some thirty to each pupa. There would not be enough room in the abode if the family were larger.

The imp is a slim and elegant creature, but oh, how small! She measures hardly two millimeters. Her garb is bronzed black, with pale legs and a heart shaped, pointed, slightly pedunculate abdomen, with never a trace of a probe for inoculating the eggs. The head is transversal, the width exceeding the length.

The male is only half the size of the female; he is also very much less numerous. Perhaps pairing is here, as we see elsewhere, a secondary matter from which it is possible to abstain, in part, without injuring the prospects of the race. Nevertheless, in the tube wherein I have housed the swarm, the few males lost among the crowd ardently woo the passing fair. There is much to be done outside, as long as the flesh fly's season lasts; things are urgent; and each pigmy hurries as fast as she can to take up her part as an exterminator.

How is the parasite's inroad into the flesh fly's pupae effected? Truth is always veiled in a certain mystery. The good fortune that secured me the ravaged pupa taught me nothing concerning the tactics of the ravager. I have never seen the Chalcidid explore the contents of my appliances; my attention was engaged elsewhere and nothing is so difficult to see as a thing not yet suspected. But, though direct observation be lacking, logic will tell us approximately what we want to know.

It is evident, to begin with, that the invasion cannot have been made through the sturdy amour of the pupae. This is too hard to be penetrated by the means at the pigmy's disposal. Naught but the delicate skin of the maggots lends itself to the introduction of the germs. An egg laying mother, therefore, appears, inspects the surface of the pool of sanies swarming with grubs, selects the one that suits her and perches on it; then, with the tip of her pointed abdomen, whence emerges, for an instant, a short probe kept hidden until then, she operates on the patient, perforating his paunch with a dexterous wound into which the germs are inserted. Probably, a number of pricks are administered, as the presence of thirty parasites seems to demand.

Anyway, the maggot's skin is pierced at either one point or many; and this happens while the grub is swimming in the pools formed by the putrid flesh. Having said this, we are faced with a question of serious interest. To set it forth necessitates a digression which seems to have nothing to do with the subject in hand and is nevertheless connected with it in the closest fashion. Without certain preliminaries, the remainder would be unintelligible. So now for the preliminaries.

I was in those days busy with the poison of the Languedocian scorpion and its action upon insects. To direct the sting toward this or the other part of the victim and moreover to regulate its emission would be absolutely impossible and also very dangerous, as long as the scorpions were allowed to act as they pleased. I wished to be able myself to choose the part to be wounded; I likewise wished to vary the dose of poison at will. How to set about it? The scorpion has no jarlike receptacle in which the venom is accumulated and stored, like that possessed, for instance, by the wasp and the bee. The last segment of the tail, gourd shaped and surmounted by the sting, contains only a powerful mass of muscles along which lie the delicate vessels that secrete the poison.

In default of a poison jar which I would have placed on one side and drawn upon at my convenience, I detach the last segment, forming the base of the sting. I obtain it from a dead and already withered scorpion. A watch glass serves as a basin. Here, I tear and crush the piece in a few drops of water and leave it to steep for four-and-twenty hours. The result is the liquid which I propose to use for the inoculation. If any poison remained in my animal's caudal gourd, there must be at least some traces of it in the infusion in the watch glass.

My hypodermic syringe is of the simplest. It consists of a little glass tube, tapering sharply at one end. By drawing in my breath, I fill it with the liquid to be tested; I expel the contents by blowing. Its point is almost as fine as a hair and enables me to regulate the dose to the degree which I want. A cubic millimeter is the usual charge. The injection has to be made at parts that are generally covered with horn. So as not to break the point of my fragile instrument, I prepare the way with a needle, with which I prick the victim at the spot required. I insert the tip of the loaded injector in the hole thus made and I blow. The thing is done in a moment, very neatly and in an orthodox fashion, favorable to delicate experiments. I am delighted with my modest apparatus.

I am equally delighted with the results. The scorpion himself, when wounding with his sting, in which the poison is not diluted as mine is in the watch glass, would not produce effects like those of my pricks. Here is something more brutal, producing more convulsion in the sufferer. The virus of my contriving excels the scorpion's.

The test is several times repeated, always with the same mixture, which, drying up by spontaneous evaporation, then made to serve again by the addition of a few drops of water, once more drained and once more moistened, does duty for an indefinite length of time. Instead of abating, the virulence increases. Moreover, the corpses of the insects operated upon undergo a curious change, unknown in my earlier observations. Then the suspicion comes to me that the actual poison of the scorpion does not enter into the matter at all. What I obtain with the end joint of the tail, with the gland at the base of the sting, I ought to obtain with any other part of the animal.

I crush in a few drops of water a joint of the tail taken from the front portion, far from the poison glands. After soaking it for twenty-four hours, I obtain a liquid whose effects are absolutely the same as those before, when I used the joint that bears the sting. I try again with the scorpion's claws, the contents of which consist solely of muscle. The results are just the same. The whole of the animal's body, therefore, no matter which fragment be submitted to the steeping process, yields the virus that so greatly pricks my curiosity.

Every part of the Spanish fly [Cantharis or blistering beetle], inside and out, is saturated with the blistering element; but there is nothing like this in the scorpion, who localizes his venom in his caudal gland and has none of it elsewhere. The cause of the effects which I observe is therefore connected with general properties which I ought to find in any insect, even the most harmless.

I consult Oryctes nasicornis, the peaceable rhinoceros beetle, on this subject. To get at the exact nature of the materials, instead of pulverizing the whole insect in a mortar, I use merely the muscular tissue obtained by scraping the inside of the dried Oryctes' corselet. Or else I extract the dry contents of the hind legs. I do the same with the desiccated corpses of the cockchafer, the Capricorn, or Cerambyx beetle, and the Cetonia, or rosechafer. Each of my gleanings, with a little water added, is left to soften for a couple of days in a watch glass and yields to the liquid whatever can be extracted from it by crushing and dissolving.

This time, we take a great step forward. All my preparations, without distinction, are horribly virulent. Let the reader judge. I select as my first patient the sacred beetle, Scarabaeus sacer, who thanks to his size and sturdiness, lends himself admirably to an experiment of this kind. I operate upon a dozen, in the corselet, on the breast, on the belly and, by preference, on one of the hind legs, far removed from the impressionable nervous centers. No matter what part my injector attacks, the effect produced is the same, or nearly. The insect falls as though struck by lightning. It lies on its back and wriggles its legs, especially the hind legs. If I set it on its feet again, I behold a sort of St. Vitus' dance. Scarabaeus lowers his head, arches his back, draws himself up on his twitching legs. He marks time with his feet on the ground, moves forward a little, moves as much backward, leans to the right, leans to the left, in wild disorder, incapable of keeping his balance or making progress. And this happens with sudden jerks and jolts, with a vigor no whit inferior to that of the animal in perfect health. It is a displacement of all the works, a storm that uproots the mutual relations of the muscles.

Seldom have I witnessed such sufferings, in my career as a cross-examiner of animals and, therefore, as a torturer. I should feel a scruple, did I not foresee that the grain of sand shifted today may one day help us by taking its place in the edifice of knowledge. Life is everywhere the same, in the Dung beetle's body as in man's. To consult it in the insect means consulting it in ourselves, means moving towards vistas which we cannot afford to neglect. That hope justifies my cruel studies, which, though apparently so puerile, are in reality worthy of serious consideration.

Of my dozen sufferers, some rapidly succumb, others linger for a few hours. They are all dead by tomorrow. I leave the corpses on the table, exposed to the air. Instead of drying and stiffening, like the asphyxiated insects intended for our collections, my patients, on the contrary, turn soft and slacken in the joints, notwithstanding the dryness of the surrounding air; they become disjointed and separate into loose pieces, which are easily removed.

The results are the same with the Capricorn, the cockchafer, the Procrustes [a large ground beetle], the Carabus [the true ground beetle, including the gold beetle]. In all of them there is a sudden break-up, followed by speedy death, a slackening of the joints and swift putrefaction. In a non-horny victim, the quick chemical changes of the tissues are even more striking. A Cetonia grub, which resists the scorpion's sting, even though repeatedly administered, dies in a very short time if I inject a tiny drop of my terrible fluid into any part of its body. Moreover, it turns very brown and, in a couple of days, becomes a mass of black putrescence.

The great peacock, that large moth who recks little of the scorpion's poison, is no more able to resist my inoculations than the sacred beetle and the others. I prick two in the belly, a male and a female. At first, they seem to bear the operation without distress. They grip the trellis work of the cage and hang without moving, as though indifferent. But soon the disease has them in its grip. What we see is not the tumultuous ending of the sacred beetle; it is the calm advent of death. With wings slackly quivering, softly they die and drop from the wires. Next day, both corpses are remarkably lax; the segments of the abdomen separate and gape at the least touch. Remove the hairs and you shall see that the skin, which was white, has turned brown and is changing to black. Corruption is quickly doing its work.

This would be a good opportunity to speak of bacteria and cultures. I shall do nothing of the sort. On the hazy borderland of the visible and the invisible, the microscope inspires me with suspicion. It so easily replaces the eye of reality by the eye of imagination; it is so ready to oblige the theorists with just what they want to see. Besides, supposing the microbe to be found, if that were possible, the question would be changed, not solved. For the problem of the collapse of the structure through the fact of a prick there would be substituted another no less obscure: how does the said microbe bring about that collapse? In what way does it go to work? Where lies its power?

Then what explanation shall I give of the facts which I have just set forth? Why, none, absolutely none, seeing that I do not know of any. As I am unable to do better, I will confine myself to a pair of comparisons or images, which may serve as a brief resting place for the mind on the dark billows of the unknown.

All of us, as children, have amused ourselves with the game of "card friars." A number of cards, as many as possible, are bent lengthwise into a semi-cylinder. They are placed on a table, one behind the other, in a winding row, the spaces in which are suitably disposed. The performance pleases the eye by its curved lines and its regular arrangement. It possesses order, which is a condition of all animated matter. You give a little tap to the first card. It falls and overturns the second, which, in the same way, topsy-turvies the third; and so on, right to the end of the row. In less than no time, the capsizing wave spreads and the handsome edifice is shattered. Order is succeeded by disorder, I might almost say, by death. What was needed thus to upset the procession of friars? A very, very slight first push, out of all proportion to the toppled mass.

Again, take a glass balloon containing a solution of alum supersaturated by heat. It is closed, during the process of boiling, with a cork and is then allowed to cool. The contents remain fluid and limpid for an indefinite period. Mobility is here represented by a faint semblance of life. Remove the cork and drop in a solid particle of alum, however infinitesimal. Suddenly, the liquid thickens into a solid lump and gives off heat. What has happened? This: crystallization has set in at the first contact of the particle of alum, the center of attraction; next, it has spread bit by bit, each solidified particle producing the solidification of those around. The impulse comes from an atom; the mass impelled is boundless. The very small has revolutionized the immense.

Of course, in the comparison between these two instances and the effects of my injections, the reader must see no more than a figure of speech, which, without explaining anything, tries to throw a glimmer of light upon it. The long procession of card friars is knocked down by the mere touch of the little finger to the first; the voluminous solution of alum suddenly turns solid under the influence of an invisible particle. In the same way, the victims of my operations succumb, thrown into convulsions by a tiny drop of insignificant size and harmless appearance.

Then what is there in that terrible liquid? First of all, there is water, inactive in itself and simply a vehicle of the active agent. If a proof were needed of its innocuousness, here is one: I inject into the thigh of any one of the sacred beetle's six legs a drop of pure water larger than that of the fatal inoculations. As soon as he is released, he makes off and trots about as nimbly as usual. He is quite firm on his legs. When put back to his pellet, he rolls it with the same zeal as before the experiment. My injection of water makes no difference to him.

What else is there in the mixture in my watch glasses? There is the disintegrated matter of the corpse, especially shreds of dried muscles. Do these substances yield certain soluble elements to water? Or are they simply reduced to a fine dust in the crushing? I will not decide this question, nor is it really of importance. The fact remains that the poison proceeds from those substances and from them alone. Animal matter, therefore, which has ceased to live is an agent of destruction within the organism. The dead cell kills the living cell; in the delicate statics of life, it is the grain of sand which, refusing its support, entails the collapse of the whole edifice.

In this connection, we may recall those dreadful dissecting room accidents. Through awkwardness, a student of anatomy pricks himself with his scalpel in the course of his work; or else, by inadvertence, he has an insignificant scratch on his hand. A cut which one would hardly notice, produced by the point of a pocket knife, a scratch of no account, from a thorn or otherwise, now becomes a mortal wound, if powerful antiseptics do not speedily remedy the ill. The scalpel is soiled by its contact with the flesh of the corpse; so are the hands. That is quite enough. The virus of corruption is introduced; and, if not treated in time, the wound proves fatal. The dead has killed the living. This also reminds us of the so-called carbuncle flies, the lancet of whose mouth parts, contaminated with the sanies of corpses, produces such terrible accidents.

My dealings as against insects are, when all is said, nothing but dissecting room wounds and carbuncle flies' stings. In addition to the gangrene that soon impairs and blackens the tissues, I obtain convulsions similar to those produced by the scorpion's sting. In its convulsive effects, the venomous fluid emitted by the sting bears a close resemblance to the muscular infusions with which I fill my injector. We are entitled, therefore, to ask ourselves if poisons, generally speaking, are not themselves a produce of demolition, a casting of the organism perpetually renewed, waste matter, in short, which, instead of being gradually expelled, is stored for purposes of attack and defense. The animal, in that case, would arm itself with its own refuse in the same way as it sometimes builds itself a home with its intestinal recrement. Nothing is wasted; life's detritus is used for self defense.

All things considered, my preparations are meat extracts. If I replace the flesh of the insect by that of another animal, the ox, for instance, shall I obtain the same results? Logic says yes; and logic is right. I dilute with a few drops of water a little Liebig's extract, that precious standby of the kitchen.

I operate with this fluid on six Cetoniae or rosechafers, four in the grub stage, two in the adult stage. At first, the patients move about as usual. Next day, the two Cetoniae are dead. The larvae resist longer and do not die until the second day. All show the same relaxed muscles, the same blackened flesh, signs of putrefaction. It is probable, therefore, that, if injected into our own veins, the same fluid would likewise prove fatal. What is excellent in the digestive tubes would be appalling in the arteries. What is food in one case is poison in the other.

A Liebig's extract of a different kind, the broth in which the liquefier puddles, is of a virulence equal, if not superior, to that of my products. All those operated upon, Capricorns, sacred beetles, ground beetles, die in convulsions. This brings us back, after a long way round, to our starting point, the maggot of the flesh fly. Can the worm, constantly floundering in the sanies of a carcass, be itself in danger of inoculation by that whereon it grows fat? I dare not rely upon experiments conducted by myself: my clumsy implements and my shaky hand make me fear that, with subjects so small and delicate, I might inflict deep wounds which of themselves would bring about death.

Fortunately, I have a collaborator of incomparable skill in the parasitic Chalcidid. Let us apply to her. To introduce her germs, she has perforated the maggot's paunch, has even done so several times over. The holes are extremely small, but the poison all around is excessively subtle and has thus been able, in certain cases, to penetrate. Now what has happened? The pupae, all from the same apparatus, are numerous. They can be divided into three not very unequal classes, according to the results supplied. Some give me the adult flesh fly, others the parasite. The rest, nearly a third, give me nothing, neither this year nor next.

In the first two cases, things have taken their normal course: the grub has developed into a fly, or else the parasite has devoured the grub. In the third case, an accident has occurred. I open the barren pupae. They are coated inside with a dark glaze, the remains of the dead maggot converted into black rottenness. The grub, therefore, has undergone inoculation by the virus through the fine openings effected by the Chalcidid. The skin has had time to harden into a shell; but it was too late, the tissues being already infected.

There you see it: in its broth of putrefaction, the worm is exposed to grave dangers. Now there is a need for maggots in this world, for maggots many and voracious, to purge the soil as quickly as possible of death's impurities. Linnaeus tells us that 'Tres muscae consumunt cadaver equi aeque cito ac leo.' [Three flies consume the carcass of a horse as quickly as a lion could do it.] There is no exaggeration about the statement. Yes, of a certainty, the offspring of the flesh fly and the bluebottle are expeditious workers. They swarm in a heap, always seeking, always snuffling with their pointed mouths. In those tumultuous crowds, mutual scratches would be inevitable if the worms, like the other flesh eaters, possessed mandibles, jaws, clippers adapted for cutting, tearing and chopping; and those scratches, poisoned by the dreadful gruel lapping them, would all be fatal.

How are the worms protected in their horrible work yard? They do not eat: they drink their fill; by means of a pepsin which they disgorge, they first turn their foodstuffs into soup; they practice a strange and exceptional art of feeding, wherein those dangerous carving implements, the scalpels with their dissecting room perils, are superfluous. Here ends, for the present, the little that I know or suspect of the maggot, the sanitary inspector in the service of the public health.

CHAPTER XVII.
RECOLLECTIONS OF CHILDHOOD

Almost as much as insects and birds—the former so dear to the child, who loves to rear his cockchafers and rose beetles on a bed of hawthorn in a box pierced with holes; the latter an irresistible temptation, with their nests and their eggs and their little ones opening tiny yellow beaks—the mushroom early won my heart with its varied shapes and colors. I can still see myself as an innocent small boy sporting my first braces and beginning to know my way through the cabalistic mazes of my reading book, I see myself in ecstasy before the first bird's nest found and the first mushroom gathered. Let us relate these grave events. Old age loves to meditate the past.

O happy days when curiosity awakens and frees us from the limbo of unconsciousness, your distant memory makes me live my best years over again. Disturbed at its siesta by some wayfarer, the partridge's young brood hastily disperses. Each pretty little ball of down scurries off and disappears in the brushwood; but, when quiet is restored, at the first summoning note they all return under the mother's wing. Even so, recalled by memory, do my recollections of childhood return, those other fledglings which have lost so many of their feathers on the brambles of life. Some, which have hardly come out of the bushes, have aching heads and tottering steps; some are missing, stifled in some dark corner of the thicket; some remain in their full freshness. Now of those which have escaped the clutches of time the liveliest are the first-born. For them the soft wax of childish memory has been converted into enduring bronze.

On that day, wealthy and leisured, with an apple for my lunch and all my time to myself, I decided to visit the brow of the neighboring hill, hitherto looked upon as the boundary of the world. Right at the top is a row of trees which, turning their backs to the wind, bend and toss about as though to uproot themselves and take to flight. How often, from the little window in my home, have I not seen them bowing their heads in stormy weather; how often have I not watched them writhing like madmen amid the snow dust which the north wind's broom raises and smoothes along the hillside! 'What are they doing up there, those desolate trees? I am interested in their supple backs, today still and upright against the blue of the sky, tomorrow shaken when the clouds pass overhead. I am gladdened by their calmness; I am distressed by their terrified gestures. They are my friends. I have them before my eyes at every hour of the day. In the morning, the sun rises behind their transparent screen and ascends in its glory. Where does it come from? I am going to climb up there and perhaps I shall find out.

I mount the slope. It is a lean grass sward close-cropped by the sheep. It has no bushes, fertile in rents and tears, for which I should have to answer on returning home, nor any rocks, the scaling of which involves like dangers; nothing but large, flat stones, scattered here and there. I. have only to go straight on, over smooth ground. But the sward is as steep as a sloping roof. It is long, ever so long; and my legs are very short. From time to time, I look up. My friends, the trees on the hilltop, seem to be no nearer. Cheerily, sonny! Scramble away!

What is this at my feet? A lovely bird has flown from its hiding place under the eaves of a big stone. Bless us, here's a nest made of hair and fine straw! It's the first I have ever found, the first of the joys which the birds are to bring me. And in this nest are six eggs, laid prettily side by side; and those eggs are a magnificent blue, as though steeped in a dye of celestial azure. Overpowered with happiness, I lie down on the grass and stare.

Meanwhile, the mother, with a little clap of her gullet—'Tack! Tack!'—flies anxiously from stone to stone, not far from the intruder. My age knows no pity, is still too barbarous to understand maternal anguish. A plan is running in my head, a plan worthy of a little beast of prey. I will come back in a fortnight and collect the nestlings before they can fly away. In the meantime, I will just take one of those pretty blue eggs, only one, as a trophy. Lest it should be crushed, I place the fragile thing on a little moss in the scoop of my hand. Let him cast a stone at me that has not, in his childhood, known the rapture of finding his first nest.

My delicate burden, which would be ruined by a false step, makes me give up the remainder of the climb. Some other day I shall see the trees on the hilltop over which the sun rises. I go down the slope again. At the bottom, I meet the parish priest's curate reading his breviary as he takes his walk. He sees me coming solemnly along, like a relic bearer; he catches sight of my hand hiding something behind my back: 'What have you there, my boy?' he asks.

All abashed, I open my hand and show my blue egg on its bed of moss.

'Ah!' says his reverence. 'A Saxicola's egg! Where did you get it?'

'Up there, father, under a stone.'

Question follows question; and my peccadillo stands confessed. By chance I found a nest which I was not looking for. There were six eggs in it. I took one of them—here it is—and I am waiting for the rest to hatch. I shall go back for the others when the young birds have their quill feathers.

'You mustn't do that, my little friend,' replies the priest. 'You mustn't rob the mother of her brood; you must respect the innocent little ones; you must let God's birds grow up and fly from the nest. They are the joy of the fields and they clear the earth of its vermin. Be a good boy, now, and don't touch the nest.'

I promise and the curate continues his walk. I come home with two good seeds cast on the fallows of my childish brain. An authoritative word has taught me that spoiling birds' nests is a bad action. I did not quite understand how the bird comes to our aid by destroying vermin, the scourge of the crops; but I felt, at the bottom of my heart, that it is wrong to afflict the mothers.

'Saxicola,' the priest had said, on seeing my find.

'Hullo!' said I to myself. 'Animals have names, just like ourselves. Who named them? What are all my different acquaintances in the woods and meadows called? What does Saxicola mean?'

Years passed and Latin taught me that Saxicola means an inhabitant of the rocks. My bird, in fact, was flying from one rocky point to the other while I lay in ecstasy before its eggs; its house, its nest, had the rim of a large stone for a roof. Further knowledge gleaned from books taught me that the lover of stony hillsides is also called the Motteux, or clodhopper, because, in the plowing season, she flies from clod to clod, inspecting the furrows rich in unearthed grubworms. Lastly, I came upon the Provencal expression Cul-blanc, which is also a picturesque term, suggesting the patch on the bird's rump which spreads out like a white butterfly flitting over the fields.

Thus did the vocabulary come into being that would one day allow me to greet by their real names the thousand actors on the stage of the fields, the thousand little flowers that smile at us from the wayside. The word which the curate had spoken without attaching the least importance to it revealed a world to me, the world of plants and animals designated by their real names. To the future must belong the task of deciphering some pages of the immense lexicon; for today I will content myself with remembering the Saxicola, or stonechat.

On the west, my village crumbles into an avalanche of garden patches, in which plums and apples ripen. Low bulging walls, blackened with the stains of lichens and mosses, support the terraces. The brook runs at the foot of the slope. It can be cleared almost everywhere at a bound. In the wider parts, flat stones standing out of the water serve as a foot bridge. There is no such thing as a whirlpool, the terror of mothers when the children are away; it is nowhere more than knee deep. Dear little brook, so tranquil, cool and clear, I have seen majestic rivers since, I have seen the boundless sea; but nothing in my memories equals your modest falls. About you clings all the hallowed pleasure of my first impressions.

A miller has bethought him of putting the brook, which used to flow so gaily through the fields, to work. Halfway up the slope, a watercourse, economizing the gradient, diverts part of the water and conducts it into a large reservoir, which supplies the mill wheels with motor power. This basin stands beside a frequented path and is walled off at the end.

One day, hoisting myself on a playfellow's shoulders, I looked over the melancholy wall, all bearded with ferns. I saw bottomless stagnant waters, covered with slimy green. In the gaps in the sticky carpet, a sort of dumpy, black-and-yellow reptile was lazily swimming. Today, I should call it a salamander; at that time, it appeared to me the offspring of the serpent and the dragon, of whom we were told such bloodcurdling tales when we sat up at night. Hoo! I've seen enough: let's get down again, quick!

The brook runs below. Alders and ash, bending forward on either bank, mingle their branches and form a verdant arch. At their feet, behind a porch of great twisted roots, are watery caverns prolonged by gloomy corridors. On the threshold of these fastnesses shimmers a glint of sunshine, cut into ovals by the leafy sieve above.

This is the haunt of the red-necktied minnows. Come along very gently, lie flat on the ground and look. What pretty little fish they are, with their scarlet throats! Clustering side by side, with their heads turned against the stream, they puff their cheeks out and in, rinsing their mouths incessantly. To keep their stationary position in the running water, they need naught but a slight quiver of their tail and of the fin on their back. A leaf falls from the tree. Whoosh! The whole troop has disappeared.

On the other side of the brook is a spinney of beeches, with smooth, straight trunks, like pillars. In their majestic, shady branches sit chattering crows, drawing from their wings old feathers replaced by new. The ground is padded with moss. At one's first step on the downy carpet, the eye is caught by a mushroom, not yet full-spread and looking like an egg dropped there by some vagrant hen. It is the first that I have picked, the first that have I turned round and round in my fingers, inquiring into its structure with that vague curiosity which is the first awakening of observation.

Soon, I find others, differing in size, shape and color. It is a real treat for my prentice eyes. Some are fashioned like bells, like extinguishers, like cups; some are drawn out into spindles, hollowed into funnels, rounded into hemispheres. I come upon some that are broken and are weeping milky tears; I step on some that, instantly, become tinged with blue; I see some big ones that are crumbling into rot and swarming with worms. Others, shaped like pears, are dry and open at the top with a round hole, a sort of chimney whence a whiff of smoke escapes when I prod their under side with my finger. These are the most curious. I fill my pockets with them to make them smoke at my leisure, until I exhaust the contents, which are at last reduced to a kind of tinder.

What fun I had in that delightful spinney! I returned to it many a time after my first find; and here, in the company of the crows, I received my first lessons in mushroom lore. My harvests, I need hardly say, were not admitted to the house. The mushroom, or the bouturel, as we called it, had a bad reputation for poisoning people. That was enough to make mother banish it from the family table. I could scarcely understand how the bouturel, so attractive in appearance, came to be so wicked; however, I accepted the experience of my elders; and no disaster ever ensued from my rash friendship with the poisoner.

As my visits to the beech clump were repeated, I managed to divide my finds into three categories. In the first, which was the most numerous, the mushroom was furnished underneath with little radiating leaves. In the second, the lower surface was lined with a thick pad pricked with hardly visible holes. In the third, it bristled with tiny spots similar to the papillae on a cat's tongue. The need of some order to assist the memory made me invent a classification for myself.

Very much later there fell into my hands certain small books from which I learnt that my three categories were well known; they even had Latin names, which fact was far from displeasing to me. Ennobled by Latin which provided me with my first exercises and translations, glorified by the ancient language which the rector used in saying his mass, the mushroom rose in my esteem. To deserve so learned an appellation, it must possess a genuine importance.

The same books told me the name of the one that had amused me so much with its smoking chimney. It is called the puffball in English, but its French name is the vesse-de-loup. I disliked the expression, which to my mind smacked of bad company. Next to it was a more decent denomination: Lycoperdon; but this was only so in appearance, for Greek roots sooner or later taught me that Lycoperdon means vesse-de-loup and nothing else. The history of plants abounds in terms which it is not always desirable to translate. Bequeathed to us by earlier ages less reticent than ours, botany has often retained the brutal frankness of words that set propriety at defiance.

How far off are those blessed times when my childish curiosity sought solitary exercise in making itself acquainted with the mushroom! 'Eheu!

Fugaces labuntur anni!' said Horace. Ah, yes, the years glide fleeting by, especially when they are nearing their end! They were the merry brook that dallies among the willows on imperceptible slopes; today, they are the torrent swirling a thousand straws along, as it rushes towards the abyss. Fleeting though they be, let us make the most of them. At nightfall, the woodcutter hastens to bind his last fagots. Even so, in my declining days, I, a humble woodcutter in the forest of science, make haste to put my bundle of sticks in order. 'What will remain of my researches on the subject of instinct? Not much, apparently; at most, one or two windows opened on a world that has not yet been explored with all the attention which it deserves.

A worse destiny awaits the mushrooms, which were my botanical joys from my earliest youth. I have never ceased to keep up my acquaintance with them. To this day, for the mere pleasure of renewing it, I go, with a halting step, to visit them on fine autumn afternoons. I still love to see the fat heads of the boletes, the tops of the agarics and the coral-red tufts of the clavaria emerge above the carpet pink with heather.

At Serignan, my last stage, they have lavished their seductions upon me, so plentiful are they on the neighboring hills, wooded with holm oak, arbutus and rosemary. During these latter years, their wealth inspired me with an insane plan: that of collecting in effigy what I was unable to keep in its natural state in an herbarium. I began to paint life size pictures of all the species in my neighborhood, from the largest to the smallest. I know nothing of the art of painting in watercolors. No matter: what I have never seen practiced I will invent, managing badly at first, then a little better, at last well. The paintbrush will make a change from the strain of my daily output of prose.

I end by possessing some hundreds of sheets representing the mushrooms of the neighborhood in their natural size and colors. My collection has a certain value. If it lacks artistic finish, at least it boasts the merit of accuracy. It brings me visitors on Sundays, country people, who stare at it in all simplicity, astounded that such fine pictures should be done by hand, without a copy and without compasses. They at once recognize the mushroom represented; they tell me its popular name, thus proving the fidelity of my brush.

Well, what will become of this great pile of drawings, the object of so much work? No doubt, my family will keep the relic for a time; but, sooner or later, taking up too much space, shifted from cupboard to cupboard, from attic to attic, gnawed by the rats, foxed, dirtied and stained, it will fall into the hands of some little grandnephews who will cut it into squares to make paper caps. It is the universal rule. What our illusions have most fondly cherished comes to a pitiful end under the claws of ruthless reality.

CHAPTER XVIII. INSECTS AND MUSHROOMS

It were out of place to recall my long relations with the bolete and the agaric if the insect did not here enter into a question of grave interest. Several mushrooms are edible, some even enjoy a great reputation; others are formidable poisons. Short of botanical studies that are not within everybody's reach, how are we to distinguish the harmless from the venomous? There is a widespread belief which says that any mushroom which insects, or, more frequently, their larvae, their grubs, accept can be accepted without fear; any mushroom which they refuse must be refused. What is wholesome food for them cannot fail to be the same for us; what is poisonous to them is bound to be equally baneful to ourselves. This is how people argue, with apparent logic, but without reflecting upon the very different capabilities of stomachs in the matter of diet. After all, may there not be some justification for the belief? That is what I purpose examining.

The insect, especially in the larval stage, is the principal devourer of the mushroom. We must distinguish between two groups of consumers. The first really eat, that is to say, they break their food into little bits, chew it and reduce it to a mouthful which is swallowed just as it is; the second drink, after first turning their food into a broth, like the bluebottles. The first are the less numerous. Confining myself to the results of my observations in the neighborhood, I count, all told, in the group of chewers, four beetles and a moth caterpillar. To these may be added the mollusk, as represented by a slug, or, more specifically, an arion, of medium size, brown and adorned with a red edge to his mantle. A modest corporation, when all is said, but active and enterprising, especially the moth.

At the head of the mushroom loving beetles, I will place a Staphylinid (Oxyporus rufus, LIN.), prettily garbed in red, blue and black. Together with his larva, which walks with the aid of a crutch at its back, he haunts the fungus of the poplar (Pholiota aegerita, FRIES). He specializes in an exclusive diet. I often come across him, both in spring and autumn, and never any elsewhere than on this mushroom. For that matter, he had made a wise choice, the epicure! This popular fungus is one of our best mushrooms, despite its color of a doubtful white, its skin which is often wrinkled and its gills soiled with rusty brown at the spores. We must not judge people by appearances, nor mushrooms either. This one, magnificent in shape and color, is poisonous; that other, so poor to look at, is excellent.

Here are two more specialist beetles, both of small size. One is the Triplax (Triplax russica, LIN.), who has an orange head and corselet and black wing-cases. His grub tackles the hispid polyporus (Polyporus hispidus, BULL.), a coarse and substantial dish, bristling at its top with stiff hairs and clinging by its side to the old trunks of mulberry trees, sometimes also of walnut and elm trees. The other is the cinnamon-colored Anisotoma (Anisotoma cinnamomea, PANZ.). His larva lives exclusively in truffles.

The most interesting of the mushroom-eating beetles is the Bolboceras (Bolboceras gallicus, MUL.). I have described elsewhere his manner of living, his little song that sounds like the chirping of a bird, his perpendicular wells sunk in search of an underground mushroom (Hydnocystis orenaria, TUL.), which constitutes his regular nourishment. He is also an ardent lover of truffles. I have taken from between his legs, at the bottom of his manor house, a real truffle the size of a hazelnut (Tuber Requienii, TUL.). I tried to rear him in order to make the acquaintance of his grub; I housed him in a large earthen pan filled with fresh sand and enclosed in a bell cover. Possessing neither hydnocistes nor truffles, I served him up sundry mushrooms of a rather firm consistency, like those of his choice. He refused them all, helvellae and clavariae, chanterelles and pezizae alike.

With a rhizopogon, a sort of little fungoid potato, which is frequent in pine woods at a moderate depth and sometimes even on the surface, I achieved complete success. I had strewn a handful of them on the sand of my breeding pan. At nightfall, I often surprised the Bolboceras issuing from his well, exploring the stretch of sand, choosing a piece not too big for his strength and gently rolling it towards his abode. He would go in again, leaving the rhizopogon, which was too large to take inside, on the threshold, where it served the purpose of a door. Next day, I found the piece gnawed, but only on the under side.

The Bolboceras does not like eating in public, in the open air; he needs the discreet retirement of his crypt. When he fails to find his food by burrowing under ground, he comes up to look for it on the surface. Meeting with a morsel to his taste, he takes it home when its size permits; if not, he leaves it on the threshold of his burrow and gnaws at it from below, without reappearing outside. Up to the present, hydnocistes, truffles and rhizopoga are the only food that I have known him to eat. These three instances tell us at any rate that the Bolboceras is not a specialist like the Oxyporus and the Triplax; he is able to vary his diet; perhaps he feeds on all the underground mushrooms indiscriminately.

The moth enlarges her domain yet further. Her caterpillar is a grub five or six millimeters long, white, with a black shiny head. Colonies of it abound in most mushrooms. It attacks by preference the top of the stem, for epicurean reasons that escape me; thence it spreads throughout the cap. It is the habitual boarder of the boletes, agarics, lactarii and russulie. Apart from certain species and certain groups, everything suits it. This puny grub, which will spin itself an infinitesimal cocoon of white silk under the piece attacked and will later become an insignificant moth, is the primordial ravager.

Let us next mention the arion, that voracious mollusk who also tackles most mushrooms of some size. He digs himself spacious niches inside them and there sits blissfully eating. Few in numbers, compared with the other devourers, he usually sets up house alone. He has, by way of a set of jaws, a powerful plane which creates great breaches in the object of his depredations. It is he whose havoc is most apparent.

Now all these gnawers can be recognized by their leavings, such as crumbs and worm holes. They dig clean passages, they slash and crumble without a slimy trail, they are the pinkers. The others, the liquefiers, are the chemists; they dissolve their food by means of reagents. All are the grubs of flies and belong to the commonalty of the Muscidae. Many are their species. To distinguish them from one another by rearing them in order to obtain the perfect stage would involve a great expenditure of time to little profit. We will describe them by the general name of maggots.

To see them at work, I select, as the field of exploitation, the satanic bolete (Boletus Satanas, LENZ.), one of the largest mushrooms that I can gather in my neighborhood. It has a dirty-white cap; the mouths of the tubes are a bright orange-red; the stem swells into a bulb with a delicate network of carmine veins. I divide a perfectly sound specimen into equal parts and place these in two deep plates, put side by side. One of the halves is left as it is: it will act as a control, a term of comparison. The other half receives on the pores of its undersurface a couple of dozen maggots taken from a second bolete in full process of decomposition.

The dissolving action of the grub asserts itself on the very day whereon these preparations are made. The undersurface, originally a bright red, turns brown and runs in every direction into a mass of dark stalactites. Soon, the flesh of the cap is attacked and, in a few days, becomes a gruel similar to liquid asphalt. It is almost as fluid as water. In this broth the maggots wallow, wriggling their bodies and, from time to time, sticking the breathing holes in their sterns above the water. It is an exact repetition of what the liquefiers of meat, the grubs of the grey flesh fly and the bluebottle, have lately shown us. As for the second half of the bolete, the half which I did not colonize with vermin, it remains compact, the same as it was at the start, except that its appearance is a little withered by evaporation. The fluidity, therefore, is really and truly the work of the grubs and of them alone.

Does this liquefaction imply an easy change? One would think so at first, on seeing how quickly it is performed by the action of the grubs. Moreover, certain mushrooms, the coprini, liquefy spontaneously and turn into a black fluid. One of them bears the expressive name of the inky mushroom (Coprinus atramentarius, BULL.) and dissolves into ink of its own accord. The conversion, in certain cases, is singularly rapid. One day, I was drawing one of our prettiest coprini (Coprinus sterquilinus, FRIES), which comes out of a little purse or volva. My work was barely done, a couple of hours after gathering the fresh mushroom, when the model had disappeared, leaving nothing but a pool of ink upon the table. Had I procrastinated ever so little, I should not have had time to finish and I should have lost a rare and interesting find.

This does not mean that the other mushrooms, especially the boletes, are of ephemeral duration and lacking in consistency. I made the attempt with the edible bolete (Boletus edulis, BULL.), the famous cepe of our kitchens, so highly esteemed for its flavor. I was wondering whether it would not be possible to obtain from it a sort of Liebig's extract of fungus, which would be useful in cooking. With this purpose, I had some of these mushrooms cut into small pieces and boiled, on the one hand, in plain water and, on the other, in water with bicarbonate of soda added. The treatment lasted two whole days. The flesh of the bolete was indomitable. To attack it, I should have had to employ violent drugs, which were inadmissible in view of the result to be attained.

What prolonged boiling and the aid of bicarbonate of soda leave almost intact the fly's grubs quickly turn into fluid, even as the flesh worms fluidify hard-boiled white of egg. This is done in each instance without violence, probably by means of a special pepsin, which is not the same in both cases. The liquefier of meat has its own brand; the liquefier of the bolete has another sort. The plate, then, is filled with a dark, running gruel, not unlike tar in appearance. If we allow evaporation free course, the broth sets, into a hard, easily crumbled slab, something like toffee. Caught in this matrix, grubs and pupa perish, incapable of freeing themselves. Analytical chemistry has proved fatal to them. The conditions are quite different when the attack is delivered on the surface of the ground. Gradually absorbed by the soil, the excess of liquid disappears, leaving the colonists free. In my dishes, it collects indefinitely, killing the inhabitants when it dries up into a solid layer.

The purple bolete (Boletus purpureus, FRIES), when subjected to the action of the maggots, gives the same result as the Satanic bolete, namely, a black gruel. Note that both mushrooms turn blue if broken and especially if crushed. With the edible bolete, whose flesh invariably remains white when cut, the product of its liquefaction by the vermin is a very pale brown. With the oronge, or imperial mushroom, the result is a broth which the eye would take for a thin apricot jam. Tests made with sundry other mushrooms confirm the rule: all, when attacked by the maggot, turn into a more or less fluid mess, which varies in color.

Why do the two boletes with the red tubes, the purple bolete and the satanic bolete, change into a dark gruel? I have an inkling of the reason. Both of them turn blue, with an admixture of green. A third species, the bluish bolete (Boletus cyanescens, BULL., var. lacteus, LEVEILLE), possess remarkable color sensitiveness. Bruise it ever so lightly, no matter where, on the cap, the stem, the tubes of the undersurface: forthwith, the wounded part, originally a pure white, is tinted a beautiful blue. Place this bolete in an atmosphere of carbonic acid gas. We can now knock it, crush it, reduce it to pulp; and the blue no longer shows. But extract a fragment from the crushed mass: immediately, at the first contact with the air, the matter turns a most glorious blue. It reminds us of a process employed in dyeing. The indigo of commerce, steeped in water containing lime and sulfate of iron, or copperas, is deprived of a part of its oxygen; it loses its color and becomes soluble in water, as it was in the original indigo plant, before the treatment which the plant underwent. A colorless liquid results. Expose a drop of this liquid to the air. Straightway, oxidization works upon the product: the indigo is reformed, insoluble and blue.

This is exactly what we see in the boletes that turn blue so readily. Could they, in fact, contain soluble, colorless indigo? One would say so, if certain properties did not give grounds for doubt. When subjected to prolonged exposure to the air, the boletes that are apt to turn blue, particularly the most remarkable, Boletus cyanescens, lose their color, instead of retaining the deep blue which would be a sign of real indigo. Be this as it may, these mushrooms contain a coloring principle which is very liable to change under the influence of the air. Why should we not regard it as the cause of the black tint when the maggots have liquefied the boletes which turn blue? The others, those with the white flesh, the edible bolete, for instance, do not assume this asphalty appearance once they are liquefied by the grubs.

All the boletes that change to blue when broken have a bad reputation; the books treat them as dangerous, or at least open to suspicion. The name of Satanic awarded to one of them is an ample proof of our fears. The caterpillar and the maggot are of another opinion: they greedily devour what we hold in dread. Now here is a strange thing: those passionate devotees of Boletus Satanas absolutely refuse certain mushrooms which we find delightful eating, including the most celebrated of all, the oronge, the imperial mushroom, which the Romans of the empire, past masters in gluttony, called the food of the gods, cibus deorum, the agaric of the Caesars, Agaricus caesareus. It is the most elegant of all our mushrooms.

When it prepares to make its appearance by lifting the fissured earth, it is a handsome ovoid formed by the outer wrapper, the volva. Then this purse gently tears and the jagged opening partly reveals a globular object of a magnificent orange. Take a hen's egg, boil it, remove the shell: what remains will be the imperial mushroom in its purse. Remove a part of the white at the top, uncovering a little of the yolk. Then you have the nascent imperial. The likeness is perfect. And so the people of my part, struck by the resemblance, call this mushroom lou rousset d'iou, or, in other words, yolk of egg. Soon, the cap emerges entirely and spreads into a disk softer than satin to the touch and richer to the eye than all the fruit of the Hesperides. Appearing amid the pink heather, it is an entrancing object.

Well, this gorgeous agaric (Amanita caesarea, SCOP.), this food of the gods the maggot absolutely refuses. My frequent examinations have never shown me an imperial attacked by the grubs in the field. It needs imprisonment in a jar and the absence of other victuals to provoke the attempt; and even then the treacle hardly seems to suit them. After the liquefaction, the grubs try to make off, showing that the fare is not to their liking. The Mollusk also, the Arion, is anything but an ardent consumer. Passing close to an imperial mushroom and finding nothing better, he stops and takes a bite, without lingering. If, therefore, we required the evidence of the insect, or even of the Slug, to know which mushrooms are good to eat, we should refuse the best of them all. Though respected by the vermin, the glorious imperial is nevertheless ruined not by larvae, but by a parasitic fungus, the Mycogone rosea, which spreads in a purply stain and turns it into a putrid mass. This is the only despoiler that I know it to possess.

A second amanita, the sheathed amanita (Amanita vaginata, BULL.), prettily streaked on the edges of the cap, is of an exquisite flavor, almost equal to the imperial. It is called lou pichot gris, the grayling, in these parts, because of its coloring, which is usually an ashen gray. Neither the maggot nor the even more enterprising Moth ever touches it. They likewise refuse the mottled amanita (Amanita pantherina, D. C.), the vernal amanita (Amanita verna, FRIES) and the lemon-yellow amanita (Amanita citrina, SCHAEFF.), all three of which are poisonous. In short, whether it be to us a delicious dish or a deadly poison, no amanita is accepted by the grubs. The arion alone sometimes bites at it. The cause of the refusal escapes us. It were vain, speaking of the mottled amanita, for instance, to allege as a reason the presence of an alkaloid fatal to the grubs, for we should have to ask ourselves why the imperial, the amanita of the Caesars, which is wholly free from poison, is rejected no less uncompromisingly than the venomous species. Could it perhaps be lack of relish, a deficiency of seasoning for stimulating the appetite? In point of fact, when eaten raw, the amanitas have no particular flavor.

What shall we learn from the sharper-flavored mushrooms? Here, in the pinewoods, is the woolly milk mushroom (Lactarius torminosus, SCHAEFF.), turned in at the edges and wrapped in a curly fleece. Its taste is biting, worse than Cayenne pepper. Torminosus means colic producing. The name is very suitable. Unless he possessed a stomach built for the purpose, the man who touched such food as this would have a singularly bad time before him. Well, that stomach the vermin possess: they revel in the pungency of the woolly milk mushroom even as the spurge caterpillar browses with delight on the loathsome leaves of the euphorbiae. As for us, we might as well, in either case, eat live coals.

Is a condiment of this kind necessary to the grubs? Not at all. Here, in the same pinewoods, is the "delicious" milk mushroom (Lactarius deliciosus, LIN.), a glorious orange-red crater, adorned with concentric zones. If bruised, it assumes a verdigris hue, possibly a variant of the indigo tint peculiar to the blue-turning boletes. From its flesh laid bare by being broken or cut ooze blood-red drops, a well-defined characteristic peculiar to this milk mushroom. Here the violent spices of the woolly milk mushroom disappear; the flesh has a pleasant taste when eaten raw. No matter: the vermin devour the mild milk mushroom with the same zest with which they devour the horribly peppered one. To them the delicate and the strong, the insipid and the peppery are all alike.

The epithet 'delicious' applied to the mushroom whose wound weeps tears of blood is highly exaggerated. It is edible, no doubt, but it is coarse eating and difficult to digest. My household refuses it for cooking purposes. We prefer to put it to soak in vinegar and afterwards to use it as we might use pickled gherkins. The real value of this mushroom is largely overrated thanks to a too laudatory epithet.

Is a certain degree of consistency required, to suit the grubs: something midway between the softness of the amanitas and the firmness of the milk mushrooms? Let us begin by questioning the olive tree agaric or luminous mushroom (Pleurotus phosphoreus, BATT.), a magnificent mushroom colored jujube red. Its popular name is not particularly appropriate. True, it frequently grows at the base of old olive trees, but I also pick it at the foot of the box, the holm oak, the plum tree, the cypress, the almond tree, the Guelder rose and other trees and shrubs. It seems fairly indifferent to the nature of the support. A more remarkable feature distinguishes it from all the other European mushrooms: it is phosphorescent. On the lower surface and there only, it sheds a soft, white gleam, similar to that of the glowworm. It lights up to celebrate its nuptials and the emission of its spores. There is no question of chemist's phosphorus here. This is a slow combustion, a sort of more active respiration than usual. The luminous emission is extinguished in the unbreathable gases, nitrogen and carbonic acid; it continues in aerated water; it ceases in water deprived of its air by boiling. It is exceedingly faint, however, so much so that it is not perceptible except in the deepest darkness. At night and even by day, if the eyes have been prepared for it by a preliminary wait in the darkness of a cellar, this agaric is a wonderful sight, looking indeed like a piece of the full moon.

Now what do the vermin do? Are they drawn by this beacon? In no wise: maggots, caterpillars and slugs never touch the resplendent mushroom. Let us not be too quick to explain this refusal by the noxious properties of the olive tree agaric, which is said to be extremely poisonous. Here, in fact, on the pebbly ground of the wastelands, is the eryngo agaric (Pleurotus eryngii, D. C.), which has the same consistency as the other. It is the berigoulo of the Provencaux, one of the most highly esteemed mushrooms. Well, the vermin will have none of it: what is a treat to us is detestable to them.

It is superfluous to continue this method of investigation: the reply would be everywhere the same. The insect, which feeds on one sort of mushroom and refuses others, cannot tell us anything about the kinds that are good or bad for us. Its stomach is not ours. It pronounces excellent what we find poisonous; it pronounces poisonous what we think excellent. That being so, when we are lacking in the botanical knowledge which most of us have neither time nor inclination to acquire, what course are we to take? The course is extremely simple.

During the thirty years and more that I have lived at Serignan, I have never heard of one case of mushroom poisoning, even the mildest, in the village; and yet there are plenty of mushrooms eaten here, especially in autumn. Not a family but, when on a walk in the mountains, gathers a precious addition to its modest alimentary resources. What do these people gather? A little of everything. Often, when rambling in the neighboring woods, I inspect the baskets of the mushroom pickers, who are delighted for me to look. I see things fit to make mycological experts stand aghast. I often find the purple bolete, which is classed among the dangerous varieties. I made the remark one day. The man carrying the basket stared at me in astonishment: 'That a poison! The wolf's bread!' he said, patting the plump bolete with his hand. 'What an idea! It's beef marrow, sir, regular beef marrow!' [Author's note: People use them indiscriminately for cooking purposes, after removing the tubes on the under side, which are easily separated from the rest of the mushroom.]

He smiled at my apprehensions and went away with a poor opinion of my knowledge in the matter of mushrooms.

In the baskets aforesaid, I find the ringed agaric (Armillaria mellea, FRIES), which is stigmatized as valde venenatus by Persoon, an expert on the subject. It is even the mushroom most frequently made use of, because of its being so plentiful, especially at the foot of the mulberry trees. I find the Satanic bolete, that dangerous tempter; the belted milk mushroom (Lactarius zonarius, BULL.), whose burning flavor rivals the pepper of its woolly kinsman; the smooth-headed amanita (Amanita leiocophala, D. C.), a magnificent white dome rising out of an ample volva and fringed at the edges with floury relics resembling flakes of casein. Its poisonous smell and soapy aftertaste should lead to suspicion of this ivory dome; but nobody seems to mind them.

How, with such careless picking, are accidents avoided? In my village and for a long way around, the rule is to blanch the mushrooms, that is to say, to bring them to the boil in water with a little salt in it. A few rinsings in cold water conclude the treatment. They are then prepared in whatever manner one pleases. In this way, what might at first be dangerous becomes harmless, because the preliminary boiling and rinsing have removed the noxious elements.

My personal experience confirms the efficacy of this rustic method. At home, we very often make use of the ringed agaric, which is reputed extremely dangerous. When rendered wholesome by the ordeal of boiling water, it becomes a dish of which I have naught but good to say. Then again the smooth-headed amanita frequently appears upon my table, after being duly boiled: if it were not first treated in this fashion, it would be hardly safe. I have tried the blue-turning boletes, especially the purple bolete and the Satanic. They answered very well to the eulogistic term of beef marrow applied to them by the mushroom picker who scouted my prudent counsels. I have sometimes employed the mottled amanita, so ill famed in the books, without disastrous result. One of my friends, a doctor, to whom I communicated my ideas about the boiling water treatment, thought that he would make the experiment on his own account. He chose the lemon-yellow amanita, which has as bad a reputation as the mottled variety, and ate it at supper. Everything went off without the slightest inconvenience. Another, a blind friend, in whose company I was one day to taste the Cossus of the Roman epicures, treated himself to the olive tree agaric, said to be so formidable. The dish was, if not excellent, at least harmless.

It results from these facts that a good preliminary boiling is the best safeguard against accidents arising from mushrooms. If the insect, devouring one species and refusing another, cannot guide us in any way, at least rustic wisdom, the fruit of long experience, prescribes a rule of conduct which is both simple and efficacious. You are tempted by a basketful of mushrooms, but you do not feel very sure as to their good or evil properties. Then have them blanched, well and thoroughly blanched. When it leaves the purgatory of the stewpan, the doubtful mushroom can be eaten without fear.

But this, you will tell me, is a system of cookery fit for savages: the treatment with boiling water will reduce the mushrooms to a mash; it will take away all their flavor and all their succulence. That is a complete mistake. The mushroom stands the ordeal exceedingly well. I have described my failure to subdue the cepes when I was trying to obtain an extract from them. Prolonged boiling, with the aid of bicarbonate of soda, so far from reducing them to a mess, left them very nearly intact. The other mushrooms whose size entitles them to culinary consideration offer the same degree of resistance. In the second place, there is no loss of succulence and hardly any of flavor. Moreover, they become much more digestible, which is a most important condition in a dish generally so heavy for the stomach. For this reason, it is the custom, in my family, to treat them one and all with boiling water, including even the glorious imperial.

I am a Philistine, it is true, a barbarian caring little for the refinements of cookery. I am not thinking of the epicure, but of the frugal man, the husbandman especially. I should consider myself amply repaid for my persistent observations if I succeeded in popularizing, however little, the wise Provencal recipe for mushrooms, an excellent food that makes a pleasant change from the dish of beans or potatoes, when we can overcome the difficulty of distinguishing between the harmless and the dangerous.

[Recorder's note: Modern mycologists warn against Fabre's claim that boiling neutralizes all mushroom poisons.]

CHAPTER XIX. A MEMORABLE LESSON

I take leave of the mushrooms with regret: there would be so many other questions to solve concerning them! Why do the maggots eat the Satanic bolete and scorn the imperial mushroom? How is it that they find delicious what we find poisonous and why is it that what seems exquisite to our taste is loathsome to theirs? Can there be special compounds in mushrooms, alkaloids, apparently, which vary according to the botanical genus? Would it be possible to isolate them and study their properties fully? Who knows whether medical science could not employ them in relieving our ailments, even as it employs quinine, morphia and other alkaloids? One might inquire into the cause of the liquefaction of the coprini, which is spontaneous, and that of the boletes, which is brought about by the maggots. Do both cases come within the same category? Does the coprinus digest itself by virtue of a pepsin similar to the maggots'? One would like to discover the oxidizable substance that gives the luminous mushroom its soft, white light, which is like the beams of the full moon. It would be interesting to know whether certain boletes turn blue owing to the presence of an indigo which is more liable to change than dyers' indigo and whether the green of the so-called delicious milk mushroom when bruised is due to a like cause.

All these patient chemical investigations would tempt me, if the rudimentary equipment of my laboratory and especially the irrevocable flight of age-worn hopes permitted it. The day has passed for it now; there is no time left to me. No matter: let us talk chemistry once more, for a little while; and, for want of something better, let us revive old memories. If the historian, now and again, takes a small place in the story of his animals, the reader will kindly excuse him: old age is prone to these reminiscences, the bloom of later days.

I have received, in all, two lessons of a scientific character in the course of my life: one in anatomy and one in chemistry. I owe the first to the learned naturalist Moquin-Tandon, who, on our return from a botanizing expedition to Monte Renoso, in Corsica, showed me the structure of a Snail in a plate filled with water. It was short and fruitful. From that moment, I was initiated. Henceforth, I was to wield the scalpel and decently to explore an animal's interior without any other guidance from a master. The second lesson, that of chemistry, was less fortunate. I will tell you what happened.

In my normal school, the scientific teaching was on an exceedingly modest scale, consisting mainly of arithmetic and odds and ends of geometry. Physics was hardly touched. We were taught a little meteorology, in a summary fashion: a word or two about a red moon, a white frost, dew, snow and wind; and, with this smattering of rustic physics, we were considered to know enough of the subject to discuss the weather with the farmer and the plowman.

Of natural history, absolutely nothing. No one thought of telling us anything about flowers and trees, which give such zest to one's aimless rambles, nor about insects, with their curious habits, nor about stones, so instructive with their fossil records. That entrancing glance through the windows of the world was refused us. Grammar was allowed to strangle life.

Chemistry was never mentioned either: that goes without saying. I knew the word, however. My casual reading, only half-understood for want of practical demonstration, had taught me that chemistry is concerned with the shuffle of matter, uniting or separating the various elements. But what a strange idea I formed of this branch of study! To me it smacked of sorcery, of alchemy and its search for the philosopher's stone. To my mind, every chemist, when at work, should have had a magic wand in his hand and the wizard's pointed, star studded cap on his head.

An important personage who sometimes visited the school, in his capacity as an honorary lecturer, was not the man to rid me of those foolish notions. He taught physics and chemistry at the grammar school. Twice a week, from eight to nine o'clock in the evening, he held a free public class in an enormous building adjacent to our schoolhouse. This was the former Church of Saint-Martial, which has today become a Protestant meeting house.

It was a wizard's cave certainly, just as I had pictured it. At the top of the steeple, a rusty weathercock creaked mournfully; in the dusk, great Bats flew all around the edifice or dived down the throats of the gargoyles; at night, Owls hooted upon the copings of the leads. It was inside, under the immensities of the vault, that my chemist used to perform. What infernal mixtures did he compound? Should I ever know?

It is the day for his visit. He comes to see us with no pointed cap: in ordinary garb, in fact, with nothing very queer about him. He bursts into our schoolroom like a hurricane. His red face is half-buried in the enormous stiff collar that digs into his ears. A few wisps of red hair adorn his temples; the top of his head shines like an old ivory ball. In a dictatorial voice and with wooden gestures, he questions two or three of the boys; after a moment's bullying, he turns on his heel and goes off in a whirlwind as he came. No, this is not the man, a capital fellow at heart, to inspire me with a pleasant idea of the things which he teaches.

Two windows of his laboratory look out upon the garden of the school. One can just lean on them; and I often come and peep in, trying to make out, in my poor brain, what chemistry can really be. Unfortunately, the room into which my eyes penetrate is not the sanctuary but a mere outhouse where the learned implements and crockery are washed. Leaden pipes with taps run down the walls; wooden vats occupy the corners. Sometimes, those vats bubble, heated by a spray of steam. A reddish powder, which looks like brick dust, is boiling in them. I learn that the simmering stuff is a dyer's root, known as madder, which will be converted into a purer and more concentrated product. This is the master's pet study.

What I saw from the two windows was not enough for me. I wanted to see farther, into the very classroom. My wish was satisfied. It was the end of the scholastic year. A stage ahead in the regular work, I had just obtained my certificate. I was free. A few weeks remain before the holidays. Shall I go and spend them out of doors, in all the gaiety of my eighteen summers? No, I will spend them at the school which, for two years past, has provided me with an untroubled roof and my daily crust. I will wait until a post is found for me. Employ my willing service as you think fit, do with me what you will: as long as I can study, I am indifferent to the rest.

The principal of the school, the soul of kindness, has grasped my passion for knowledge. He encourages me in my determination; he proposes to make me renew my acquaintance with Horace and Virgil, so long since forgotten. He knows Latin, he does; he will rekindle the dead spark by making me translate a few passages. He does more: he lends me an Imitation with parallel texts in Latin and Greek. With the first text, which I am almost able to read, I will puzzle out the second and thus increase the small vocabulary which I acquired in the days when I was translating Aesop's Fables. It will be all the better for my future studies. What luck! Board and lodging, ancient poetry, the classical languages, all the good things at once!

I did better still. Our science master—the real, not the honorary one—who came twice a week to discourse of the rule of three and the properties of the triangle, had the brilliant idea of letting us celebrate the end of the school year with a feast of learning. He promised to show us oxygen. As a colleague of the chemist in the grammar school, he obtained leave to take us to the famous laboratory and there to handle the object of his lesson under our very eyes. Oxygen, yes, oxygen, the all-consuming gas; that was what we were to see on the morrow. I could not sleep all night for thinking of it.

Thursday afternoon came at last. As soon as the chemistry lesson is over, we were to go for a walk to Les Angles, the pretty village over yonder, perched on a steep rock. We were therefore in our Sunday best, our out-of-doors clothes: black frock coats and tall hats. The whole school was there, some thirty of us, in the charge of an usher, who knew as little as we did of the things which we were about to see. We crossed the threshold of the laboratory, not without excitement. I entered a great nave with a Gothic roof, an old, bare church through which one's voice echoed, into which the light penetrated discreetly through stained glass windows set in ribs and rosettes of stone. At the back were huge raised benches, with room for an audience of many hundreds; at the other end, where the choir once was, stood an enormous chimney mantel; in the middle was a large, massive table, corroded by the chemicals. At one end of this table was a tarred tub, lined inside with lead and filled with water. This, I at once learned, was the pneumatic trough, the vessel in which the gases were collected.

The professor begins the experiment. He takes a sort of large, long glass bulb, bent abruptly in the region of the neck. This, he informs us, is a retort. He pours into it, from a screw of paper, some black stuff that looks like powdered charcoal. This is manganese dioxide, the master tells us. It contains in abundance, in a condensed state and retained by combination with the metal, the gas which we propose to obtain. An oily looking liquid, sulfuric acid, an excessively powerful agent, will set it at liberty. Thus filled, the retort is placed on a lighted stove. A glass tube brings it into communication with a bell jar full of water on the shelf of the pneumatic trough. Those are all the preparations. What will be the result? We must wait for the action of heat.

My fellow pupils gather eagerly round the apparatus, cannot come close enough to it. Some of them play the part of the fly on the wheel and glory in contributing to the success of the experiment. They straighten the retort, which is leaning to one side; they blow with their mouths on the coals in the stove. I do not care for these familiarities with the unknown. The good natured master raises no objection; but I have never been able to endure the thronging of a crowd of gapers, who are very busy with their elbows and force their way to the front row to see whatever is happening, even though it be merely a couple of mongrels fighting. Let us withdraw and leave these officious ones to themselves. There is so much to see here, while the oxygen is being prepared. Let us make the most of the occasion and take a look round the chemist's arsenal.

Under the spacious chimney mantel is a collection of queer stoves, bound round with bands of sheet iron. There are long and short ones, high and low ones, all pierced with little windows that are closed with a terracotta shutter. This one, a sort of little tower, is formed of several parts placed one above the other and each supplied with big round handles to hold them by when you take the monument to pieces. A dome, with an iron chimney, tops the whole edifice, which must be capable of producing a very hell fire to roast a stone of no significance. Another, a squat one, stretches out like a curved spine. It has a round hole at either end; and a thick porcelain tube sticks out from each. It is impossible to conceive the purpose which such instruments as these can serve. The seekers of the philosopher's stone must have had many like them. They are torturers' engines, tearing the metals' secrets from them.

The glass things are arranged on shelves. I see retorts of different sizes, all with necks bent at a sudden angle. In addition to their long beak, some of them have a narrow little tube coming out of their bulb. Look, youngster, and do not try to guess the object of these curious vessels. I see glasses with feet to them, funnel-shaped and deep; I stand amazed at strange looking bottles with two or three mouths to each, at phials swelling into a balloon with a long, narrow tube. What an odd array of implements! And here are glass cupboards with a host of bottles and jars, filled with all manner of chemicals. The labels apprise me of their contents: molybdenite of ammonia, chloride of antimony, permanganate of potash and ever so many other strange terms. Never, in all my reading, have I met with such repellent language.

Suddenly, bang! And there is running and stamping and shouting and cries of pain! What has happened? I rush up from the back of the room. The retort has burst, squirting its boiling vitriol in every direction. The wall opposite is all stained with it. Most of my fellow pupils have been more or less struck. One poor youth has had the splashes full in his face, right into his eyes. He is yelling like a madman. With the help of a friend who has come off better than the others, I drag him outside by main force, take him to the sink, which fortunately is close at hand, and hold his face under the tap. This swift ablution serves its purpose. The horrible pain begins to be allayed, so much so that the sufferer recovers his senses and is able to continue the washing process for himself.

My prompt aid certainly saved his sight. A week later, with the help of the doctor's lotions, all danger was over. How lucky it was that I took it into my head to keep some way off! My isolation, as I stood looking into the glass case of chemicals, left me all my presence of mind, all my readiness of resource. What are the others doing, those who got splashed through standing too near the chemical bomb? I return to the lecture hall. It is not a cheerful spectacle. The master has come off badly: his shirtfront, waistcoat and trousers are covered with smears, which are all smoldering and burning into holes. He hurriedly divests himself of a portion of his dangerous raiment. Those of us who possess the smartest clothes lend him something to put on so that he can go home decently.

One of the tall, funnel-shaped glasses which I was admiring just now is standing, full of ammonia, on the table. All, coughing and sniveling, dip their handkerchiefs into it and rub the moist rag over their hats and coats. In this way, the red stains left by the horrible compound are made to disappear. A drop of ink will presently restore the color completely.

And the oxygen? There was no more question, I need hardly say, of that. The feast of learning was over. Never mind: the disastrous lesson was a mighty event for me. I had been inside the chemist's laboratory; I had had a glimpse of those wonderful jars and tubes. In teaching, what matters most is not the thing taught, whether well or badly grasped: it is the stimulus given to the pupil's latent aptitudes; it is the fulminate awakening the slumbering explosives. One day, I shall obtain on my own account that oxygen which ill luck has denied me; one day, without a master, I shall yet learn chemistry.

Yes, I shall learn this chemistry, which started so disastrously. And how? By teaching it. I do not recommend that method to anybody. Happy the man who is guided by a master's word and example! He has a smooth and easy road before him, lying straight ahead. The other follows a rugged path, in which his feet often stumble; he goes groping into the unknown and loses his way. To recover the right road, if want of success have not discouraged him, he can rely only on perseverance, the sole compass of the poor. Such was my fate. I taught myself by teaching others, by passing on to them the modicum of seed that had ripened on the barren moor cleared, from day to day, by my patient plowshare.

A few months after the incident of the vitriol bomb, I was sent to Carpentras to take charge of junior classes at the college there. The first year was a difficult one, swamped as I was by the excessive number of pupils, a set of duffers kept out of the more advanced classes and all at different stages in spelling and grammar. Next year, my school is divided into two; I have an assistant. A weeding-out takes place in my crowd of scatterbrains. I keep the older, the more intelligent ones; the others are to have a term in the preparatory division. From that day forward, things are different. Curriculum there is none. In those happy times, the master's personality counted for something; there was no such thing as the scholastic piston working with the regularity of a machine. It was left for me to act as I thought fit. Well, what should I do to make the school earn its title of 'upper primary'?

Why, of course! Among other things, I shall do some chemistry! My reading has taught me that it does no harm to know a little chemistry, if you would make your furrows yield a good return. Many of my pupils come from the country; they will go back to it to improve their land. Let us show them what the soil is made of and what the plant feeds on. Others will follow industrial careers; they will become tanners, metal founders, distillers; they will sell cakes of soap and kegs of anchovies. Let us show them pickling, soap making, stills, tannin and metals. Of course, I know nothing about these things, but I shall learn, all the more so as I shall have to teach them to the boys; and your schoolboy is a little demon for jeering at the master's hesitation.

As it happens, the college boasts a small laboratory, containing just what is strictly indispensable: a receiver, a dozen glass balloons, a few tubes and a niggardly assortment of chemicals. That will do, if I can have the run of it. But the laboratory is a sanctum reserved for the use of the sixth form. No one sets foot in it except the professor and his pupils preparing for their degree. For me, the outsider, to enter that tabernacle with my band of young imps would be most unseemly; the rightful occupant would never think of allowing it. I feel it myself: elementary teaching dare not aspire to such familiarity with the higher culture. Very well, we will not go there, so long as they will lend me the things.

I confide my plan to the principal, the supreme dispenser of those riches. He is a classics man, knows hardly anything of science, at that time held in no great esteem, and he does not quite understand the object of my request. I humbly insist and exert my powers of persuasion. I discreetly emphasize the real point of the matter. My group of pupils is a numerous one. It takes more meals at the schoolhouse—the real concern of a principal—than any other section of the college. This group must be encouraged, lured on, increased if possible. The prospect of disposing of a few more platefuls of soup wins the battle for me; my request is granted. Poor science! All that diplomacy to gain your entrance among the despised ones, who have not been nourished on Cicero and Demosthenes!

I am authorized to move, once a week, the material required for my ambitious plans. From the first floor, the sacred dwelling of the scientific things, I shall take them down to a sort of cellar where I give my lessons. The troublesome part is the pneumatic trough. It has to be emptied before it is carried downstairs and to be filled again afterwards. A day scholar, a zealous acolyte, hurries over his dinner and comes to lend me a hand an hour or two before the class begins. We effect the move between us.

What I am after is oxygen, the gas which I once saw fail so lamentably. I thought it all out at my leisure, with the help of a book. I will do this, I will do that, I will go to work in this or the other fashion. Above all, we will run no risks, perhaps of blinding ourselves; for it is once more a question of heating manganese dioxide with sulfuric acid. I am filled with misgivings at the recollection of my old school fellow yelling like mad. Who cares? Let us try for all that: fortune favors the brave! Besides, we will make one prudent condition, from which I shall never depart: no one but myself shall come near the table. If an accident happen, I shall be the only one to suffer; and, in my opinion, it is worth a burn or two to make acquaintance with oxygen.

Two o'clock strikes; and my pupils enter the classroom. I purposely exaggerate the likelihood of danger. They are all to stay on their benches and not stir. This is agreed. I have plenty of elbow room. There is no one by me, except my acolyte, standing by my side, ready to help me when the time comes. The others look on in profound silence, reverent towards the unknown.

Soon the gaseous bubbles come "gloo-glooing" through the water in the bell jar. Can it be my gas? My heart beats with excitement. Can I have succeeded without any trouble at the first attempt? We will see. A candle blown out that moment and still retaining a red tip to its wick is lowered by a wire into a small test jar filled with my product. Capital! The candle lights with a little explosion and burns with extraordinary brilliancy. It is oxygen right enough.

The moment is a solemn one. My audience is astounded and so am I, but more at my own success than at the relighted candle. A puff of vainglory rises to my brow; I feel the fire of enthusiasm run through my veins. But I say nothing of these inner sensations. Before the boys' eyes, the master must appear an old hand at the things he teaches. What would the young rascals think of me if I allowed them to suspect my surprise, if they knew that I myself am beholding the marvelous subject of my demonstration for the first time in my life? I should lose their confidence, I should sink to the level of a mere pupil.

Sursum corda! Let us go on as if chemistry were a familiar thing to me. It is the turn of the steel ribbon, an old watch spring rolled corkscrew fashion and furnished with a bit of tinder. With this simple lighted bait, the steel should take fire in a jar filled with my gas. And it does burn; it becomes a splendid firework, with cracklings and a blaze of sparks and a cloud of rust that tarnishes the jar. From the end of the fiery coil a red drop breaks off at intervals, shoots quivering through the layer of water left at the bottom of the vessel and embeds itself in the glass which has suddenly grown soft. This metallic tear, with its indomitable heat, makes every one of us shudder. All stamp and cheer and applaud. The timid ones place their hands before their faces and dare not look except through their fingers. My audience exults; and I myself triumph. Ha, my friends, isn't it grand, this chemistry!

All of us have red letter days in our lives. Some, the practical men, have been successful in business; they have made money and hold their heads high in consequence. Others, the thinkers, have gained ideas; they have opened a new account in the ledger of nature and they silently taste the hallowed joys of truth. One of my great days was that of my first acquaintance with oxygen. On that day, when my class was over and all the materials put back in their place, I felt myself grow several inches taller. An untrained workman, I had shown, with complete success, that which was unknown to me a couple of hours before. No accident whatever, not even the least stain of acid.

It is, therefore, not so difficult nor so dangerous as the pitiful finish of the Saint Martial lesson might have led me to believe. With a vigilant eye and a little prudence, I shall be able to continue. The prospect is enchanting.

And so, in due season, comes hydrogen, carefully contemplated in my reading, seen and reseen with the eye of the mind before being seen with the eyes of the body. I delight my little rascals by making the hydrogen flame sing in a glass tube, which trickles with the drops of water resulting from the combustion; I make them jump with the explosions of the thunderous mixture. Later, I show them, with the same invariable success, the splendors of phosphorus, the violent powers of chlorine, the loathsome smells of sulfur, the metamorphoses of carbon and so on. In short, in a series of lessons, the principal nonmetallic elements and their compounds are passed in review during the course of the year.

The thing was bruited abroad. Fresh pupils came to me, attracted by the marvels of the school. Additional places were laid in the dining hall; and the principal, who was more interested in the profits on his beans and bacon than in chemistry, congratulated me on this accession of boarders. I was fairly started. Time and an indomitable will would do the rest.

CHAPTER XX. INDUSTRIAL CHEMISTRY

Everything happens sooner or later. When, through the low windows overlooking the garden of the school, my eye glanced at the laboratory, where the madder vats were steaming; when, in the sanctuary itself, I was present, by way of a first and last chemistry lesson, at the explosion of the retort of sulfuric acid that nearly disfigured every one of us, I was far indeed from suspecting the part which I was destined to play under that same vaulted roof. Had a prophet foretold that I should one day succeed the master, never would I have believed him. Time works these surprises for us.

Stones would have theirs too, if anything were able to astonish them. The Saint Martial building was originally a church; it is a protestant place of worship now. Men used to pray there in Latin; today they pray in French. In the intervening period, it was for some years in the service of science, the noble orison that dispels the darkness. What has the future in store for it? Like many another in the ringing city, to use Rabelais' epithet, will it become a home for the fuller's teasels, a warehouse for scrap iron, a carrier's stable? Who knows? Stones have their destinies no less unexpected than ours.

When I took possession of it as a laboratory for the municipal course of lectures, the nave remained as it was at the time of my former short and disastrous visit. To the right, on the wall, a number of black stains struck the eye. It was as though a madman's hand, armed with the inkpot, had smashed its fragile projectile at that spot. I recognized the stains at once. They were the marks of the corrosive which the retort had splashed at our heads. Since those days of long ago, no one had thought fit to hide them under a coat of whitewash. So much the better: they will serve me as excellent counselors. Always before my eyes, at every lesson, they will speak to me incessantly of prudence.

For all its attractions, however, chemistry did not make me forget a long cherished plan well suited to my tastes, that of teaching natural history at a university. Now, one day, at the grammar school, I had a visit from a chief inspector which was not of an encouraging nature. My colleagues used to call him the Crocodile. Perhaps he had given them a rough time in the course of his inspections. For all his boorish ways, he was an excellent man at heart. I owe him for a piece of advice which greatly influenced my future studies.

That day, he suddenly appeared, alone, in the schoolroom, where I was taking a class in geometrical drawing. I must explain that, at this time, to eke out my ridiculous salary and, at all costs, to provide a living for myself and my large family, I was a mighty pluralist, both inside the college and out. At the college in particular, after two hours of physics, chemistry or natural history, came, without respite, another two hours' lesson, in which I taught the boys how to make a projection in descriptive geometry, how to draw a geodetic plane, a curve of any kind whose law of generation is known to us. This was called graphics.

The sudden irruption of the dread personage causes me no great flurry. Twelve o'clock strikes, the pupils go out and we are left alone. I know him to be a geometrician. The transcendental curve, perfectly drawn, may work upon his gentler mood. I happen to have in my portfolio the very thing to please him. Fortune serves me well in this special circumstance. Among my boys, there is one who, though a regular dunce at everything else, is a first rate hand with the square, the compass and the drawing pen: a deft-fingered numskull, in short.

With the aid of a system of tangents of which I first showed him the rule and the method of construction, my artist has obtained the ordinary cycloid, followed by the interior and the exterior epicycloid and, lastly, the same curves both lengthened and shortened. His drawings are admirable Spider's webs, encircling the cunning curve in their net. The draftsmanship is so accurate that it is easy to deduce from it beautiful theorems, which would be very laborious to work out by the calculus.

I submit the geometrical masterpieces to my chief inspector, who is himself said to be smitten with geometry. I modestly describe the method of construction, I call his attention to the fine deductions which the drawing enables one to make. It is labor lost: he gives but a heedless glance at my sheets and flings each on the table as I hand it to him.

'Alas!' said I to myself. 'There is a storm brewing; the cycloid won't save you; it's your turn for a bite from the Crocodile!'

Not a bit of it. Behold the bugbear growing genial. He sits down on a bench, with one leg here, another there, invites me to take a seat by his side and, in a moment, we are discussing graphics. Then, bluntly: 'Have you any money?' he asks.

Astounded at this strange question, I answer with a smile.

'Don't be afraid,' he says. 'Confide in me. I'm asking you in your own interest. Have you any capital?'

'I have no reason to be ashamed of my poverty, monsieur l'inspecteur general. I frankly admit, I possess nothing; my means are limited to my modest salary.'

A frown greets my answer; and I hear, spoken in an undertone, as though my confessor were talking to himself: 'That's sad, that's really very sad.'

Astonished to find my penury treated as sad, I ask for an explanation: I was not accustomed to this solicitude on the part of my superiors.

'Why, yes, it's a great pity,' continues the man reputed so terrible. 'I have read your articles in the Annales des sciences naturelles. You have an observant mind, a taste for research, a lively style and a ready pen. You would have made a capital university professor.'

'But that's just what I'm aiming at!'

'Give up the idea.'

'Haven't I the necessary attainment?'

'Yes, you have; but you have no capital.' The great obstacle stands revealed to me: woe to the poor in pocket! University teaching demands a private income. Be as ordinary, as commonplace as you please, but, above all, possess the coin that lets you cut a dash. That is the main thing; the rest is a secondary condition.

And the worthy man tells me what poverty in a frock coat means. Though less of a pauper than I, he has known the mortification of it; he describes it to me, excitedly, in all its bitterness. I listen to him with an aching heart; I see the refuge which was to shelter my future crumbling before my eyes: 'You have done me a great service, sir,' I answered. 'You put an end to my hesitation. For the moment, I give up my plan. I will first see if it is possible to earn the small fortune which I shall need if I am to teach in a decent manner.'

Thereupon we exchanged a friendly grip of the hand and parted. I never saw him again. His fatherly arguments had soon convinced me: I was prepared to hear the blunt truth. A few months earlier, I had received my nomination as an assistant lecturer in zoology at the university of Poitiers. They offered me a ridiculous salary. After paying the costs of moving, I should have had hardly three francs a day left; and, on this income, I had to keep my family, numbering seven in all. I hastened to decline the very great honor.

No, science ought not to practice these jests. If we humble persons are of use to her, she should at least enable us to live. If she can't do that, then let her leave us to break stones on the highway. Oh, yes, I was prepared for the truth when that honest fellow talked to me of frock coated poverty! I am telling the story of a not very distant past. Since then, things have improved considerably; but, when the pear was properly ripened, I was no longer of an age to pick it.

And what was I to do now, to overcome the difficulty mentioned by my inspector and confirmed by my personal experience? I would take up industrial chemistry. The municipal lectures at Saint Martial placed a spacious and fairly well-equipped laboratory at my disposal. Why not make the most of it?

The chief manufacture of Avignon was madder. The farmer supplied the raw material to the factories, where it was turned into purer and more concentrated products. My predecessor had gone in for it and done well by it, so people said. I would follow in his footsteps and use the vats and furnaces, the expensive plant which I had inherited. So to work.

What should I set myself to produce? I proposed to extract the coloring substance, alizarin, to separate it from the other matters found with it in the root, to obtain it in the pure state and in a form that allowed of the direct printing of the stuffs, a much quicker and more artistic method than the old dyeing process.

Nothing could be simpler than this problem, once the solution was known; but how tremendously obscure while it had still to be solved! I dare not call to mind all the imagination and patience spent upon endless endeavors which nothing, not even the madness of them, discouraged. What mighty meditations in the somber church! What glowing dreams, soon to be followed by sore disappointment, when experiment spoke the last word and upset the scaffolding of my plans. Stubborn as the slave of old amassing a peculium for his enfranchisement, I used to reply to the check of yesterday by the fresh attempt of tomorrow, often as faulty as the others, sometimes the richer by an improvement, and I went on indefatigably, for I too cherished the indomitable ambition to set myself free.

Should I succeed? Perhaps so. I at last had a satisfactory answer. I obtained, in a cheap and practical fashion, the pure coloring matter, concentrated in a small volume and excellent for both printing and dyeing. One of my friends took up my process on a large scale in his works; a few calico factories adopted the produce and expressed themselves delighted with it. The future smiled at last; a pink rift opened in my gray sky. I should possess the modest fortune without which I must deny myself the pleasure of teaching in a university. Freed of the torturing anxiety about my daily bread, I should be able to live at ease among my insects.

In the midst of the joys of seeing these problems solved by chemistry, yet another ray of sunshine was reserved for me, adding its gladness to that of my success. Let us go back a couple of years. The chief inspectors visited our grammar school. These personages travel in pairs: one attends to literature, the other to science. When the inspection was over and the books checked, the staff was summoned to the principal's drawing room, to receive the parting admonitions of the two luminaries. The man of science began. I should be sadly put to it to remember what he said. It was cold professional prose, made up of soulless words which the hearer forgot once the speaker's back was turned, words merely boring to both. I had heard enough of these chilly sermons in my time; one more of them could not hope to make an impression on me.

The inspector in literature spoke next. At the first words which he uttered, I said to myself: 'Oho! This is a very different business!'

The speech was alive and vigorous and full of images; indifferent to scholastic commonplaces, the ideas soared, hovering gently in the serene heights of a kindly philosophy. This time, I listened with pleasure; I even felt stirred. Here was no official homily: it was full of impassioned zeal, of words that carried you with them, uttered by an honest man accomplished in the art of speaking, an orator in the true sense of the word. In all my school experience, I had never had such a treat.

When the meeting broke up, my heart beat faster than usual: 'What a pity,' I thought, 'that my side, the science side, cannot bring me into contact, some day, with that inspector! It seems to me that we should become great friends.'

I inquired his name of my colleagues, who were always better informed than I. They told me it was Victor Duruy.

Well, one day, two years later, as I was looking after my Saint Martial laboratory in the midst of the steam from my vats, with my hands the color of boiled lobster claws from constant dipping in the indelible red of my dyes, there walked in, unexpectedly, a person whose features straightway seemed familiar. I was right, it was the very man, the chief inspector whose speech had once stirred me. M. Duruy was now minister of public instruction. He was styled, 'Your excellency;' and this style, usually an empty formula, was well deserved in the present case, for our new minister excelled in his exalted functions. We all held him in high esteem. He was the workers' minister, the man for the humble toiler.

'I want to spend my last half-hour at Avignon with you,' said my visitor, with a smile. 'That will be a relief from the official bowing and scraping.'

Overcome by the honor paid me, I apologized for my costume—I was in my shirt sleeves—and especially for my lobster claws, which I had tried, for a moment, to hide behind my back.

'You have nothing to apologize for. I came to see the worker. The working man never looks better than in his overall, with the marks of his trade on him. Let us have a talk. What are you doing just now?'

I explained, in a few words, the object of my researches; I showed my product; I executed under the minister's eyes a little attempt at printing in madder red. The success of the experiment and the simplicity of my apparatus, in which an evaporating dish, maintained at boiling point under a glass funnel, took the place of a steam chamber, caused him some surprise.

'I will help you,' he said. 'What do you want for your laboratory?'

'Why, nothing, monsieur le ministre, nothing! With a little application, the plant I have is ample.'

'What, nothing! You are unique there! The others overwhelm me with requests; their laboratories are never well enough supplied. And you, poor as you are, refuse my offers!'

'No, there is one thing which I will accept.'

'What is that?'

'The signal honor of shaking you by the hand.'

'There you are, my friend, with all my heart. But that's not enough. What else do you want?'

'The Paris Jardin des Plantes is under your control. Should a crocodile die, let them keep the hide for me. I will stuff it with straw and hang it from the ceiling. Thus adorned, my workshop will rival the wizard's cave.'

The minister cast his eyes round the nave and glanced up at the Gothic vault: 'Yes, it would look very well.' And he gave a laugh at my sally. 'I now know you as a chemist,' he continued. 'I knew you already as a naturalist and a writer. I have heard about your little animals. I am sorry that I shall have to leave without seeing them. They must wait for another occasion. My train will be starting presently. Walk with me to the station, will you? We shall be alone and we can chat a bit more on the way.'

We strolled along, discussing entomology and madder. My shyness had disappeared. The self sufficiency of a fool would have left me dumb; the fine frankness of a lofty mind put me at my ease. I told him of my experiments in natural history, of my plans for a professorship, of my fight with harsh fate, my hopes and fears. He encouraged me, spoke to me of a better future. We reached the station and walked up and down outside, talking away delightfully.

A poor old woman passed, all in rags, her back bent by age and years of work in the fields. She furtively put out her hand for alms. Duruy felt in his waistcoat, found a two franc piece and placed it in the outstretched hand; I wanted to add a couple of sous as my contribution, but my pockets were empty, as usual. I went to the beggar woman and whispered in her ear: 'Do you know who gave you that? It's the emperor's minister.

The poor woman started; and her astounded eyes wandered from the open-handed swell to the piece of silver and from the piece of silver to the open-handed swell. What a surprise! What a windfall!

'Que lou bon Dieu ie done longo vido e santa, pecaire!' she said, in her cracked voice.

And, curtseying and nodding, she withdrew, still staring at the coin in the palm of her hand.

'What did she say?' asked Duruy.

'She wished you long life and health.' 'And pecaire?'

'Pecaire is a poem in itself: it sums up all the gentler passions.'

And I myself mentally repeated the artless vow. The man who stops so kindly when a beggar puts out her hand has something better in his soul than the mere qualities that go to make a minister.

We entered the station, still alone, as promised, and I quite without misgivings. Had I but foreseen what was going to happen, how I should have hastened to take my leave! Little by little, a group formed in front of us. It was too late to fly; I had to screw up my courage. Came the general of division and his officers, came the prefect and his secretary, the mayor and his deputy, the school inspector and the pick of the staff. The minister faced the ceremonial semicircle. I stood next to him. A crowd on one side, we two on the other. Followed the regulation spinal contortions, the empty obeisances which my dear Duruy had come to my laboratory to forget. When bowing to St. Roch, in his corner niche, the worshipper at the same time salutes the saint's humble companion. I was something like St. Roch's dog in the presence of those honors which did not concern me. I stood and looked on, with my awful red hands concealed behind my back, under the broad brim of my felt hat.

After the official compliments had been exchanged, the conversation began to languish; and the minister seized my right hand and gently drew it from the mysterious recesses of my wide awake.

'Why don't you show those gentlemen your hands?' he said. 'Most people would be proud of them.'

'Workman's hands,' said the prefect's secretary. 'Regular workman's hands.'

The general, almost scandalized at seeing me in such distinguished company, added: 'Hands of a dyer and cleaner.'

'Yes, workman's hands,' retorted the minister, 'and I wish you many like them. Believe me, they will do much to help the chief industry of your city. Skilled as they are in chemical work, they are equally capable of wielding the pen, the pencil, the scalpel and the lens. As you here seem unaware of it, I am delighted to inform you.'

This time, I should have liked the ground to open and swallow me up. Fortunately, the bell rang for the train to start. I said goodbye to the minister and, hurriedly taking to flight, left him laughing at the trick which he had played me.

The incident was noised about, could not help being so, for the peristyle of a railway station keeps no secrets. I then learned to what annoyances the shadow of the great exposes us. I was looked upon as an influential person, having the favor of the gods at my disposal. Place hunters and canvassers tormented me. One wanted a license to sell tobacco and stamps, another a scholarship for his son, another an increase of his pension. I had only to ask and I should obtain, said they.

O simple people, what an illusion was yours! You could not have hit upon a worse intermediary. I figuring as a postulant! I have many faults, I admit, but that is certainly not one of them. I got rid of the importunate people as best I could, though they were utterly unable to fathom my reserve. What would they have said had they known of the minister's offers with regard to my laboratory and my jesting reply, in which I asked for a crocodile skin to hang from my ceiling! They would have taken me for an idiot.

Six months elapsed; and I received a letter summoning me to call upon the minister at his office. I suspected a proposal to promote me to a more important grammar school and wrote begging that I might be left where I was, among my vats and my insects. A second letter arrived, more pressing than the first and signed by the minister's own hand. This letter said: 'Come at once, or I shall send my gendarmes to fetch you.'

There was no way out of it. Twenty-four hours later, I was in M. Duruy's room. He welcomed me with exquisite cordiality, gave me his hand and, taking up a number of the Moniteur: 'Read that,' he said. 'You refused my chemical apparatus; but you won't refuse this.

I looked at the line to which his finger pointed. I read my name in the list of the Legion of Honor. Quite stupid with surprise, I stammered the first words of thanks that entered my head.

'Come here,' said he, 'and let me give you the accolade. I will be your sponsor. You will like the ceremony all the better if it is held in private, between you and me: I know you!'

He pinned the red ribbon to my coat, kissed me on both cheeks, made me telegraph the great event to my family. What a morning, spent with that good man!

I well know the vanity of decorative ribbonry and tinware, especially when, as too often happens, intrigue degrades the honor conferred; but, coming as it did, that bit of ribbon is precious to me. It is a relic, not an object for show. I keep it religiously in a drawer.

There was a parcel of big books on the table a collection of the reports on the progress of science drawn up for the International Exhibition of 1867, which had just closed.

'Those books are for you,' continued the minister. 'Take them with you. You can look through them at your leisure: they may interest you. There is something about your insects in them. You're to have this too: it will pay for your journey. The trip which I made you take must not be at your own expense. If there is anything over, spend it on your laboratory.'

And he handed me a roll of twelve hundred francs. In vain I refused, remarking that my journey was not so burdensome as all that; besides, his embrace and his bit of ribbon were of inestimable value compared with my disbursements. He insisted: 'Take it,' he said, 'or I shall be very angry. There's something else: you must come to the emperor's with me tomorrow, to the reception of the learned societies.'

Seeing me greatly perplexed and as though demoralized by the prospect of an imperial interview: 'Don't try to escape me,' he said, 'or look out for the gendarmes of my letter! You saw the fellows in the bearskin caps on your way up. Mind you don't fall into their hands. In any case, lest you should be tempted to run away, we will go to the Tuileries together, in my carriage.'

Things happened as he wished. The next day, in the minister's company, I was ushered into a little drawing room at the Tuileries by chamberlains in knee breeches and silver-buckled shoes. They were queer people to look at. Their uniforms and their stiff gait gave them the appearance, in my eyes, of beetles who, by way of wing cases, wore a great, gold-laced dress coat, with a key in the small of the back. There were already a score of persons from all parts waiting in the room. These included geographical explorers, botanists, geologists, antiquaries, archeologists, collectors of prehistoric flints, in short, the usual representatives of provincial scientific life.

The emperor entered, very simply dressed, with no parade about him beyond a wide, red, watered silk ribbon across his chest. No sign of majesty, an ordinary man, round and plump, with a large moustache and a pair of half-closed, drowsy eyelids. He moved from one to the other, talking to each of us for a moment as the minister mentioned our names and the nature of our occupations. He showed a fair amount of information as he changed his subject from the ice floes of Spitzbergen to the dunes of Gascony, from a Carlovingian charter to the flora of the Sahara, from the progress in beetroot growing to Caesar's trenches before Alesia. When my turn came, he questioned me upon the hypermetamorphosis of the Meloidae [a beetle family including the oil beetle and the Spanish fly], my last essay in entomology. I answered as best I could, floundering a little in the proper mode of address, mixing up the everyday monsieur with sire, a word whose use was so entirely new to me. I passed through the dread straits and others succeeded me. My five minutes' conversation with an imperial majesty was, they tell me, a most distinguished honor. I am quite ready to believe them, but I never had a desire to repeat it.

The reception came to an end, bows were exchanged and we were dismissed. A luncheon awaited us at the minister's house. I sat on his right, not a little embarrassed by the privilege; on his left was a physiologist of great renown. Like the others, I spoke of all manner of things, including even Avignon Bridge. Duruy's son, sitting opposite me, chaffed me pleasantly about the famous bridge on which everybody dances; he smiled at my impatience to get back to the thyme-scented hills and the gray olive yards rich in Grasshoppers.

'What!' said his father. 'Won't you visit our museums, our collections? There are some very interesting things there.'

'I know, monsieur le ministre, but I shall find better things, things more to my taste, in the incomparable museum of the fields.'

'Then what do you propose to do?'

'I propose to go back tomorrow.

I did go back, I had had enough of Paris: never had I felt such tortures of loneliness as in that immense whirl of humanity. To get away, to get away was my one idea.

Once home among my family, I felt a mighty load off my mind and a great joy in my heart, where rang a peal of bells proclaiming the delights of my approaching emancipation. Little by little, the factory that was to set me free rose skywards, full of promises. Yes, I should possess the modest income which would crown my ambition by allowing me to descant on animals and plants in a university chair.

'Well, no,' said Fate, 'you shall not acquire the freedman's peculium; you shall remain a slave, dragging your chain behind you; your peal of bells rings false!'

Hardly was the factory in full swing when a piece of news was bruited, at first a vague rumor, an echo of probabilities rather than certainties, and then a positive statement leaving no room for doubt. Chemistry had obtained the madder dye by artificial means; thanks to a laboratory concoction, it was utterly overthrowing the agriculture and industries of my district. This result, while destroying my work and my hopes, did not surprise me unduly. I myself had toyed with the problem of artificial alizarin and I knew enough about it to foresee that, in no very distant future, the work of the chemist's retort would take the place of the work of the fields.

It was finished; my hopes were dashed to the ground. What to do next? Let us change our lever and begin to roll Sisyphus' stone once more. Let us try to draw from the ink pot what the madder vat declines to yield. Laboremus!

CPSIA information can be obtained
at www.ICGtesting.com
Printed in the USA
LVHW041002020323
740706LV00008B/812

9 789356 899063